IMPACT OF NATURAL HAZARDS ON OIL AND GAS EXTRACTION:
THE SOUTH CASPIAN BASIN

E. Bagirov
Conoco Inc.
P.O. Box 2197
Houston, Texas

and

I. Lerche
Department of Geological Sciences
University of South Carolina
Columbia, South Carolina

Kluwer Academic / Plenum Publishers
New York, Boston, Dordrecht, London, Moscow

Library of Congress Cataloging-in-Publication Data

Lerche, I. (Ian)
    Impact of natural hazards on oil and gas extraction: the South Caspian Basin / Ian
Lerche and Elchin Bagirov
        p. cm.
    Includes bibliographical references and index.
    ISBN 0-306-46285-0
    1. Petroleum--Geology--Caspian Sea region. 2. Hazardous geographic
environments--Caspian Sea Region. 3. Natural disasters--Caspian Sea Region. I.
Bagirov, Elchin. II. Title

TN875 .L47 2000
622'.338'09475--dc21   99-047693

ISBN 0-306-46285-0

©1999 Kluwer Academic/Plenum Publishers, New York
233 Spring Street, New York, N.Y. 10013

http://www.wkap.com

10 9 8 7 6 5 4 3 2 1

Printed in the United States of America

PREFACE

Since the dissolution of the Soviet Union almost a decade ago, there has been rapid evolution of interactions between the Western nations and individual countries of the former Soviet Union. As part of that interaction, the autonomous independent Republic of Azerbaijan through its scientific arm, the Geological Institute of the Azerbaijan Academy of Sciences under the Directorship of Academician Akif Ali-Zadeh and Deputy Director Ibrahim Guliev, arranged for personnel to be seconded to the University of South Carolina. The idea here was to see to what extent a quantitative understanding could be achieved of the evolution of the Azerbaijan part of the South Caspian Basin from dynamical, thermal and hydrocarbon perspectives. The Azeris brought with them copious amounts of data collected over decades which, together with the quantitative numerical codes available at USC, enabled a concerted effort to be put forward, culminating in two large books (Evolution of the South Caspian Basin: Geological Risks and Probable Hazards, 675 pps; and The South Caspian Basin: Stratigraphy, Geochemistry, and Risk Analysis, 472 pps.) both of which were published by the Azerbaijan Academy of Sciences, and also many scientific papers. Thus, over the last four to five years an integrated comprehensive start has been made to understand the hydrocarbon proneness of the South Caspian Basin.

In the course of the endeavor to understand the basinal evolution, it became clear that a variety of natural hazards occur in the Basin. These hazards are important for their potential impact on rigs, sub-sea completion equipment, pipelines, and for drilling hazards. Over the time of the basin evaluation study, we have also been compiling different hazard assessments based both on historical data bases and on quantitative hazard techniques constrained by geological, chemical and physical conditions appropriate for the onshore and offshore components of the South Caspian Basin.

The purpose of this monograph is to provide technical details of the hazards so that some appreciation is available of the likely worst case conditions for each hazard and their frequencies of occurrence. In each case we tried not just to estimate the possibility of the event, but to answer the question "what will happen if the event does indeed occur?" In this way one can plan strategies and tactics for rig-siting, pipelines, sub-sea completions, and for drilling. While not necessarily eliminating potential hazards, the ideas and estimates presented here may at least minimize the influence of particular hazards. All examples are taken from one single basin - the South Caspian - to show the variation of natural hazards in one geological setting, in order to compare the scales of possible damages caused by such phenomena within a consistent geological framework.

In our opinion, the work presented here should be readily understood by the average graduate student, and should be of considerable use to those individuals in the oil industry whose job it is to assess practical hazards as their companies undertake hydrocarbon exploration and/or production efforts. We believe the tools and techniques proposed in this book can be used by such individuals in their assessments of natural hazards. Companies working in the South Caspian region can use directly the specific results presented in the book.

We are particularly grateful to Akif Ali-Zadeh, Ibrahim Guliev, and their many colleagues at the Geological Institute of Azerbaijan for their unending cooperation. We thank the Industrial Associates of the Basin Modeling Group at USC for their financial support. And we are most appreciative of families and friends who suffer the most during the many stages of writing a book. This work would not have been possible without the unfailing support of our secretary, Donna Black, who, once again, converted hieroglyphic handwriting to legible English typescript.

To you all, and many others, we extend our thanks.

Elchin Bagirov
Ian Lerche
Columbia, SC

CONTENTS

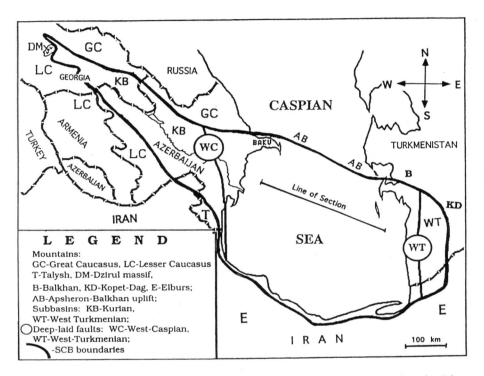

Location Map 1. Major Features of the South Caspian Basin, showing the Line of Section of the 12 second seismic line.

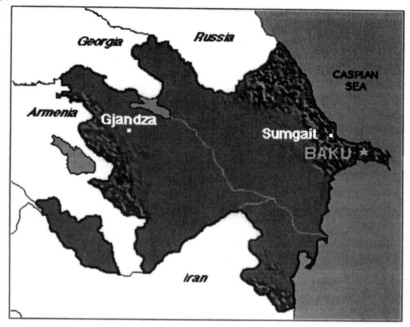

Location Map 2. Expanded view of Azerbaijan and its relation to neighboring countries.

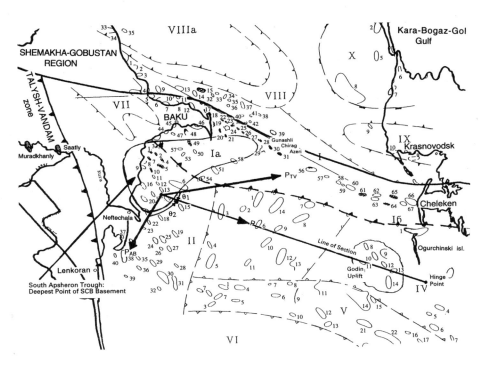

Location Map 3. The known offshore anticlinal regions (marked as closed, roughly elliptical contours) together with thrusting directions from the Apsheron-Balkhan region (P_{AB}) and from the Talysh-Vandam region (P_{TV}). Also marked are principal geological areas.

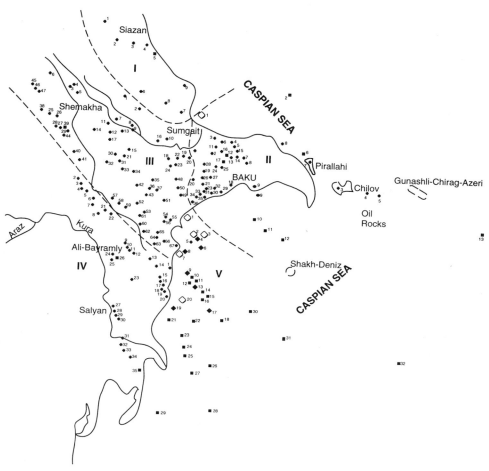

Location Map 4. Map of Mud Volcanoes of Azerbaijain. This map is followed
by a listing of all mud volcanoes known in the region.

LOCATION MAPS OF GEOLOGICAL STRUCTURES IN THE SOUTH CASPIAN BASIN

I. PRISCASPIAN REGION

1. Kaynarja
2. Saadan
3. Nardaran
4. Khydyrzyndy
5. Zorat
6. Khanaga
7. Kurkachidag
8. Kohna-Gady
9. Shorabad

II. APSHERON REGION

A. Apsheron Peninsula

1. Arbat (Agdag)
2. Atabatar Sparkling Springs
3. Girvaalty
4. Pilpila (Sisai)
5. Chullutepe
6. Gullutepe
7. Kirmaku
8. Bog-Boga
9. Zykh
10. Sangar
11. Kemuratan
12. Abikh
13. Keireki
14. Beyukdag
15. Kichikdag
16. Kechaldag
17. Zigilpiri
18. Bibi-Eibat
19. Damlamaja
20. Shorbulag
21. Gulbakht
22. Saryncha
23. Shongar
24. Bozdag Kobi
25. Bozdag Gekmaly
26. West Bozdag
27. Bozdag Guzdek
28. Uchtepe
29. Lokbatan

24. Kirkishlag
25. Madrasa
26. Sarabil
27. Charagan
28. Nugedi
29. Melikchobanly
30. Shorsulu
31. Sheytanud
32. Ayazakhtarma
33. Nardaranakhtarma
34. Suleyman
35. Cheilakhtarma
36. Cheildag
37. Davalidag
38. Sagiyan
39. Sabirli
40. Kushchi
41. Kelany
42. Ajiveli
43. Kalendarakhtarma
44. Jenikend
45. Zeyva
46. Sarysura
47. Bizlan
48. Agzygyr
49. Shikigaya
50. Chapylmysh
51. Ultagi
52. Arzani
53. Touragai
54. Bolshoi Kanizadag
55. Goturlug
56. Malyi Kanizadag
57. Dashmardan
58. Shokikhan
59. Durandag
60. Solakhai
61. Agdam
62. Ayrantekyan
63. Goturdag
64. Delaniz
65. Geyarchin
66. Dashgil
67. Bahar (Alyat Cape)

IV. KURIAN REGION

1. Hamamdag

2. Injabel (North)
3. Injabel (South)
4. Kalamaddin
5. Shorbachi
6. Zaakhtarma
7. Akhtarma-Pashaly
8. Malyi Kharami
9. Gyzdag
10. Ekizdag
11. Malyi Mishovdag
12. Bolshoi Mishovdag
13. Kalmas
14. Khydyrly
15. Agzybir A
16. Agzybir B
17. Agzybir C
18. Aralyg
19. Dovshan dagy
20. Bandovan
21. Bolshoi Kharami
22. Gyrlyg
23. Kursangi
24. Pirgarin
25. Gektepe
26. Yandere
27. Babazanan
28. Oil Salse
29. Gyrlyg Lake
30. Durovdag
31. Tatarmahla
32. Duzdag
33. Kichik Pilpilya
34. Neftchala
35. Yenigyshlag

V. BAKU ARCHIPELAGO

1. Sangachal
2. Yama Submarine Crater
3. North Duvanny Crater
4. Duvanny Island
5. Alyat-More
6. Bulla Island
7. Alyat Dome
8. Glinyanny Island
9. Garasu (Los) Island
10. Savenkova Bank
11. Bank (10 feet)

1 INTRODUCTION TO THE QUANTITATIVE ASSESSMENT OF GEOLOGICAL HAZARDS

Commercial oil and gas accumulations form in specific geological settings. Commonly those geological conditions are associated with different sorts of geological hazards. In addition, exploration and production equipment is very expensive. An ever-increasing percentage of oil reserves are concentrated offshore, which requires construction of special deep-water platforms, vessels, pipelines, etc. At the same time damage to such facilities can result in losing expensive construction, and may also lead to massive spillage of oil into the sea water or onto land. And then remediation costs can be much higher than all operational expenses and capital costs. The evacuation of personnel from platforms also takes a longer time than for onshore rigs where people can move themselves, using vehicles, or simply run away. Therefore, all oil companies pay much attention to the hazard aspects of the areas of future work and, before any kind of activity, they start with sea bed monitoring, shallow seismic and acquisition studies in order to estimate probable hazard activities and the most hazardous zones in their license areas. However, as a rule those investigations show "static" conditions of hazards in the area: microseismicity, positions of mud diapirs, gas pockets, hydrate layers, gryphons, etc. It is of great importance to provide "dynamic" estimates of <u>how often</u> a phenomenon can occur, <u>how strong</u> it can be, and <u>how far</u> its influence can extend in a most likely worst case. There are two approaches to the quantitative assessment of those questions: statistical analysis of historical data and computer modeling the processes. (Lab experiments or physical modeling of such phenomena like earthquakes or mud eruptions are impossible to simulate on true-scale miniatures).

The first approach is very good in cases where one has a representative set of observations. For instance, in the earthquake case, we used a group of seismic station observations over tens of years to estimate the seismic risk for a specific area. At the same time, for mud volcanic eruption hazards the statistical analysis was not so simple, because of omissions in observed data, and those missing data were not random because the probability of being missed depended highly on the strength of eruption. Thus for statistical estimation of such events one has to keep in mind that fact and consider the probability of missing data during hazard assessment analysis.

1

Some observations are completely absent, or there are very limited data available. For instance, there are available measurements of the mud flow sizes for onshore mud volcanoes, but not for offshore submarine flows. And the dynamics of mud or sediment flows for air and submarine conditions are completely different, so one cannot use onshore data to estimate the offshore mud flows. In such a case, dynamic modeling of mud flows in water conditions is recommended. The same approach was used for other phenomena for which very limited data were available, like hydrate hazards, gas hazards, horizontal stress hazards. A very effective procedure is to combine the dynamic modeling with the limited observed data (overpressure, breccia and flame hazards). The only curious, and very rare, phenomena which we could not assess quantitatively is the appearance and disappearance of mud islands related to the vertical motion of mud diapirs crests. But the scattered observations of mud island behavior were collected here from different sources to concentrate the attention of industry on the possibility of such potentially very hazardous events. At the end of the volume, economical aspects of hazards are shown, indicating how to handle the risk for hazard situations in hydrocarbon exploration and production risk assessments for profitability.

Of course, to compare the scales of different hazards it is interesting to show them relative to the same geological background - an example of one single basin. In our case we chose the South Caspian Basin, which is a place of concentration of an enormous variety of different geological hazards. This basin is a young, rapidly subsiding basin. Hydrocarbon exploration and production companies operating in such basins can expect similar hazards to abound; while for different geological conditions the dominance of individual types of hazards can be completely different. But, using the approaches given in the book, one can make hazard estimations. First we provide some information on the geology of the South Caspian Basin, where all our examples are taken from.

B. Some Information on the South Caspian Basin

1. Historical Oil Review

The South Caspian Basin (SCB) covers the onshore and offshore regions of Azerbaijan, Iran, Georgia and Turkmenistan (Fig. 1.1). Conventionally and geographically the SCB is divided into three parts: (i) the western part is usually called the Kurian sub-basin, limited from the north by the Greater Caucasus mountains, from the southwest by the Lesser Caucasus and by Talysh, and from the east by the coast line between onshore and the Caspian Sea, offshore Azerbaijan. The eastern part of the Kurian sub-basin extends as a narrow wedge into Georgian territory up to the Dzirul massif; (ii) the eastern part of the SCB is usually labeled the West-Turkmenian sub-basin, limited by Kopet-Dag, the Balkhans and the coast line of the Caspian Sea along Turkmenistan and Iran; (iii) the central part of the SCB is situated between the

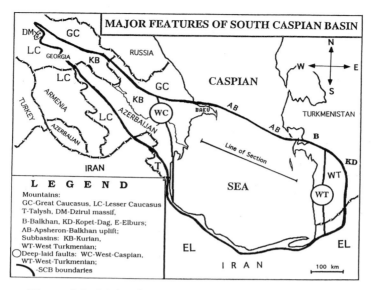

Figure 1.1. Major features of the South Caspian Basin.

Kurian and West-Turkmenian sub-basins, with the Apsheron-Balkhan uplift as the northern border and the Elburs Mountains in Iran as the southern border.

The region is known as the oldest oil province in the world. In ancient times crude oil was extracted from shallow holes, which were dug with spades. The extraction was also processed by hand and later using animals. Oil was produced in the Apsheron Peninsula and later in the areas of the lower Kura deep and from the Balkhan zone of the West Turkmenian deep. At the beginning of the 19th century holes were dug not only onshore, but also offshore Caspian Sea. In 1848 the first real well was drilled in the Bibi-Eibat field. But the revolution in the oil industry in the region came in 1874 with a special law of the Russian Tsar allowing one to buy and use productive lands. After that time large numbers of domestic and foreign businessmen invested in the oil fields. The names of Rothschild and Nobel are mentioned here as some of the richest people. The fast development of the industry required new scientific applications. Among them were the first log borehole measurements, made by the Schlumberger brothers. Nobel built the first refinery plant and also the first tanker in the world, which was called "Zoroastr", the name of the holy prophet of the Zoroastrism religion, which was spread in the southern part of Azerbaijan before Islam came and was connected with hundreds of natural burning hydrocarbon seeps.

At the end of the last century the bay near the Bibi-Eibat field was sand-filled and, on the new artificial land, more wells were drilled. In this way the development of the first offshore field started. In 1925 the first wood platform was constructed and in 1934 the first metal platform was built on the Pirallahi (Artem) island to the north of the Apsheron Peninsula.

For a long time the region was almost the only crude oil base of the Soviet Union. During the Second World War Azerbaijan provided more than 90% of the oil production of the Soviet Union. Many new fields were then discovered and explored, mostly in the Apsheron Peninsula. After the war exploration efforts moved to the offshore, to the Kura Basin, and to the Turkmenian part of the Basin.

To the present time more than 1.2 billion tons of crude oil and condensate have been produced in the region. The region is now the focus of interest of western oil companies, who immediately started activity in Azerbaijan, Turkmenistan and Georgia when the communist system collapsed in those countries and free business development was possible. Presumably the same sort of activity will develop in Iran as the political situation warrants.

But, associated with this world-class oil and gas province are major natural hazards which can seriously influence exploration and production operations.

Many of the hazards can be associated with dynamically evolving mud diapirs and volcanoes (the South Caspian Basin has over 300 such volcanoes, about 40% of the total known worldwide). Such hazards include flaming eruptions, mud flows, breccia ejecta, hydrate formation, and gas emissions, and occur both onshore and offshore.

Other hazards include earthquakes (both crustal and in the sedimentary pile), together with faults and fractures often in association with mud diapir activity.

Turbidites, mud diapir slumping and mud diapir lateral motion, recorded appearances and disappearances of offshore mud islands - often very quickly - also indicate an extremely dynamic regime which can contribute to the rapid development of hazards on a very short timescale.

In addition, high sedimentary overpressure and recorded hydrate observations for offshore mud diapirs provide direct drilling hazards.

It would seem appropriate to investigate at least some of the orders of magnitude of these various phenomena so that some idea is to hand of the likely potential for encountering one or more such problems while exploring for hydrocarbons or producing oil and/or gas.

Each chapter of this monograph investigates individual hazards using statistical data where possible and also quantitative model representations to provide an overview of likely ranges of conditions contributing to hazards for different aspects of rigs, infrastructure equipment, and personnel.

However, to better understand the nature of these hazards, a geological overview is now provided with emphasis on the offshore part of the basin, where the geological hazard is high and where almost all attention of oil companies is concentrated at the moment.

2. Tectonical Review

The SCB contains sediments from Jurassic to Quaternary in age. Discovered oil and gas deposits are sited in reservoir traps ranging in age from Upper Cretaceous to Quaternary.

Formation of the SCB is connected with closure of the Tethys Ocean. Since the end of Mesozoic time, collision of micro-continents, island arcs, active margins and fragments of ocean crust have taken place in this region. Parts of the ocean crust, heavier than neighboring fragments of transitional type crust (island arcs), gradually flexed and subsided under the action of lateral compressive thrust. Similar Tethyian oceanic crust relics are observed directly in the Mediterranean, Black Sea, and South Caspian, and are sometimes called "oceanic windows". In contrast to other "oceanic windows" inherited from Tethys, the basement of the SCB has already subsided to a depth of 25-30 km and has been covered by massive sediment input. At present, according to geophysical data (Bagir-Zadeh et al., 1988), the epicenter of the deepest position of the SCB basement is situated approximately 70-80 km to the south of Baku, Azerbaijan and the shallowest position is situated in the area of the Godin uplift (Fig. 1.2).

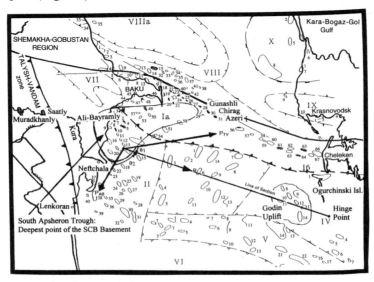

Figure 1.2. Major features of the central part of SCB, small elliptical areas are known anticlinal regions.

In the west of the SCB, in the central part of the Kurian sub-basin, is the zone of buried volcanic uplifts, with the Talysh-Vandam gravitation maximum being a reflection of this zone. The discovery of the Muradkhanly oil field in weathered volcanic rocks (basic and calc-alkanic composition) was the stimulus for drilling other volcanic uplifts of this zone. The drilling of an ultra-deep well on the Saatly uplift allowed a study of the deepest section of these volcanic rocks (to 8300 m depth). As a result of this study, made during the 1980's, the idea that these volcanic uplifts have an island arc genesis was placed on a firm basis (Moshashvili, 1982; Nadirov, 1983, 1985; Salakhov, 1985; Kremenetzky, 1988).

The zone of volcanic uplift is limited from the east by the so-called "West-Caspian deep-laid fault" (Fig. 1.1). Azerbaijani researchers have taken "the deep-laid fault" as a border, dividing the Kurian sub-basin into Lower Kurian and Middle Kurian parts. The deep-laid fault is not seen on the surface nor in well-sections drilled along the fault line to the comparatively shallow depths of 3-5 km. However, the deep-laid fault appears throughout all different types of geophysical investigations (gravitational, magnetic, and different types of seismic). On the Saatly structure, which is situated only 10-15 km to the west of the line of the fault, middle-Jurassic volcanic rocks were penetrated around 2700m (Salakhov, 1985). The geophysical border, which is taken as the whole Jurassic surface, is interpreted to be at a depth of 11-13 km (Bagir-Zadeh et al., 1988) to the east of the West-Caspian deep-laid fault (on its eastern down-thrown block). It is unlikely that such a large fault offset can be formed at just one time, i.e., instantaneously; it would seem that the fault developed progressively as a growth fault with sedimentation.

Nadirov (1985) suggested that the zone of contrast of geophysical changes, which traditionally is taken as representing the "West Caspian deep-laid fault", constitutes a sub-vertical, butt-ended (paving block), front margin of the island arc fragment. The subsided block of the fault is interpreted as a buried fragment of the accretional complex associated with the front arc, which is situated shallower than the subductional zone. In this case, the Mesozoic floor of Gobustan through the South Caspian and the Baku Archipelago is interpreted as an accretionary complex located above the subducted ocean crust. Based on the current disposition of structural elements from fragments of the island arc, viewed as relics of ocean crust, this subduction can be characterized as directed to the west and, in all probability, existed from Middle Jurassic time and continued to the end of Upper Cretaceous time.

From the beginning of the Paleogene period both volcanic processes associated with subduction of the island arc fragment, as well as the subduction process itself, decreased in activity and both of these processes appear to have finally terminated in Eocene time.

Since Oligocene time, the Kurian-South Caspian zone of the crust was subject to the influence of orogenic processes which developed in the Greater and Lesser Caucasus and in Talysh. The approach of an active margin from the north, associated with the orogenic process of the Greater Caucasus evolution, involved the north edge of the South Caspian ocean fragment in the process of northward-driven subduction.

During Neogene time the orogenic processes around the South Caspian fragment increased. The Kurian sub-basin is most obviously subject to the influence of orogenic processes of the Greater and Lesser Caucasus Mountains. In Azerbaijan, along the northern border of the SCB, undulations of anticline folds are parallel to sub-parallel with the Greater Caucasus direction. In the Shamakha-Gobustan region the development of prolonged anticline folds is observed, with the ratio of fold length to width reaching five-fold and greater, and many of the brakhi-anticline folds are broken by faults paralleling the Greater Caucasus. In the same region, the development has taken place of imbricated thrust covers, the offset of which reaches around 10 km. All of these events point to a pressure from the north margin of the basin. The Apsheron-Balkhan threshold step, which is the prolongation of the Greater Caucasus to the east, divided the South Caspian into the shallow Middle Caspian Basin and the comparatively deep water South Caspian Basin (bathymetry to 1000 m).

Therefore, in the Azerbaijan part of the SCB, two main directions of pressure took place: (i) Western - in the direction to the West Caspian deep-laid fault (or in the direction to the Talysh-Vandam buried uplift zone); and (ii) Sub-northern - in the direction to the Greater Caucasus - Apsheron-Balkhan prolongation.

The processes of creation of folds play important roles in the genesis of oil and gas traps, primarily for anticline structures. Grouping the existent anticline structures according to their directions provides an indication of the direction of total effective pressure, the reason for the anticlinal structure formations.

As an example, in the Lower Kurian sub-basin the axes of oil and gas field structures have directions parallel to Talysh-Vandam, while in the Shamakha-Gobustan area the corresponding direction is sub-parallel to that of the Greater Caucasus. The Apsheron Peninsula is subject to the influence of both of these pressure directions. In the Baku Archipelago region the effect of partial superposition (vectorial addition) of the pressure in both directions is observed. The axes of the anticline zones passing through the structures radiate in a fan-shaped manner between both borders in the process of extension from the Lower Kurian sub-basin to the deep-water, central part of the South Caspian.

The directions of the anticlines located in the Apsheron-Balkhan uplift zone are mainly parallel with the direction of the uplift zone. Nevertheless, the axis passing through the Guneshli-Chirag-Azeri major oil and gas fields is rotated in the same direction as the "fan" of the Baku Archipelago anticline zones. The implication is that, in spite of the relatively large distance from the anticline zone, the Guneshli-Chirag-Azeri fields were all subject to some pressure influence coming from the west (Talysh-Vandam) direction during the process of fold creation.

The southern anticline structures of the Iranian offshore have directions parallel to sub-parallel with the Elburs Mountains. In this part of the SCB the fold formation processes are connected with compression associated with the influence of the Elburs Mountains orogenesis.

In the Turkmenian part of the SCB, it is possible to identify two structure field groups in accordance with their spatial location as well as with their axes directions. In the northern (Kopet-Dag) part, the structural orientation is mainly subordinated to the Apsheron-Balkhan direction. Further to the south, field areas (mainly in the onshore) have a meridian (N-S) direction of their fold axes, which coincides with the direction of the so-called "West-Turkmenian-threshold". Ideas concerning the nature of this threshold vary from the deep-laid fault to the flexural fold. In the central part of the SCB the folds have relatively isometric shapes. Nevertheless, their meridian orientations and prolongations are clearly visible.

Therefore, around the South Caspian the axes of the anticline zones are oriented relatively parallel to the edges of the basin and/or to the deep-laid faults of the West Caspian (in Azerbaijan) and West Turkmenia.

Remarkable is the fact that just that part of the SCB situated between the two "deep-laid faults" underwent the most flexure and subsidence, with the region enjoying the unique properties of the thickest sediment cover, lowest thermal gradient, mud diapirism, mud volcanism, and native oceanic crust basement. Sometimes just this large central part is taken as the SCB in whole. The behavior of the anticline zones described above shows the complicated circular disposition of the compressions, which are directed from the edges of the SCB to its central part.

All of these facts imply that the oceanic fragment edges (the SCB basement) had similar flexural subduction during Mesozoic-Cenozoic times under the forces of the over-riding active neighboring plates. Greater surety of subduction to the West (under Talysh-Vandam) and to the North (under the Greater Caucasus-Apsheron-Balkhan Peninsula) directions can be guaranteed because one further confirmation is that, in accordance with the geophysical data, the deepest point of the SCB basement is located 50-70 km south of Baku (Fig. 1.2), directly in the wedge between the two directions. This deepest point of the SCB basement may be a "triple junction point", i.e. the point of crossing of the north and west subduction relict troughs.

One of the more immediate problems of quantitative basin analysis of the SCB is the determination of the way stress from each of the different directions influenced basement evolution and sediment cover development.

3. Sedimentation rates of the South Caspian Basin

One of the unique properties of the SCB is the thick sediment cover, which reaches 30 km in some places. The existence of such a cover implies a corresponding basement subsidence. The subsidence and the corresponding sedimentation took place irregularly with geological time in the SCB.

As a base for quantitative analysis of basin subsidence and sedimentation rates, a 12-second seismic cross-section was used from the Azerbaijan shelf (in the west) to the Turkmenian shelf (in the east) and passing through the central deep-water part of the SCB (Gambarov et al., 1993). Eight seismic-stratigraphic units (SSU), confined to different age sediments of the SCB, are picked out on this regional cross-section. The boundaries between these SSU are schematically shown in fig. 1.3. A part of the 12-second seismic cross-section is shown on figure 1.4.

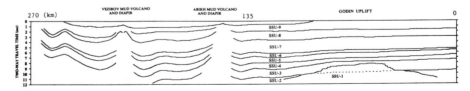

Figure 1.3. The seismic-stratigraphic unit boundaries of the regional cross-section through the central part of SCB.

The maximum thickness of sediment cover can be seen from the sharpness of the boundaries on the cross-section, and can be observed in shallow parts of the continental slopes of both the Azerbaijanian and Turkmenian offshore regions,

Figure 1.4. Mud diapir on part of the 12-second two-way-travel cross-section.

as well as in the central deep-water part of the SCB (conventionally called "abyssal").

Diagrams of sedimentation rate (on a logarithmic scale) vs. geological time were developed for each continental slope and for the abyssal part of the SCB separately. Average values of the SCB's sedimentation rates during geological time were then computed from:

$$\bar{R}_j = \bar{H}_j/dT_j \tag{1.1}$$

and

$$\bar{H}_j = \sum_{i=1}^{n} S_i / \sum_{i=1}^{n} I_i = \sum_{i=1}^{n} (h_i + h_{i-1})/2L \tag{1.2}$$

where $\bar{R}_j$ is the average sedimentation rate; $\bar{H}_j$ is the average thickness; dT_j the time duration; L the total section length; S_j is the section area between two current thicknesses; h_i is the current thickness along the section length.

As is clear from the diagrams of figure 1.5, beginning from middle Pliocene time, the sedimentation rates underwent a rapid increase of approximately one order of magnitude and reached turbidite (avalanche) values. As shown in Table 1.1, sedimentation rates for the middle Pliocene reached 1.35km/My for the Azerbaijan slope, exceeded 1.4 km/Myr for the Turkmenian slope, were at 1.3 km/My in the central abyssal part of the SCB, and the average value for the SCB was about 1.4 km/Myr.

10

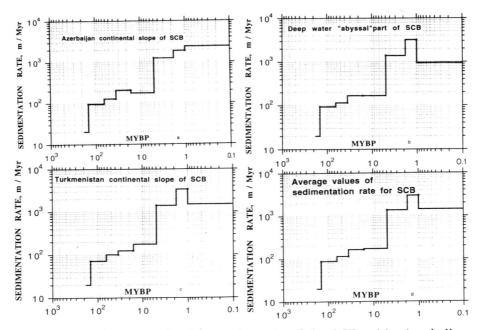

Figure 1.5. Diagrams of sedimentation rate of the SCB: (a) Azerbaijan continental slope; (b) Central deep-water "abyssal" part; (c) Turkmenistan continental slope; (d) average values for SCB.

During the Quaternary, the sedimentation rate continues to increase only in the Azerbaijan part of the SCB, reaching 2.4 km/Myr. In other parts of the SCB the sedimentation rates show some decrease after their maximum peaks in Upper Pliocene time. These results are shown graphically in Fig. 1.5a,b,c,d and also in Table 1.1. The sedimentation rate values for Jurassic sediments, given in Table 1.1, have not been taken from SCB data, but according to current oceanic abyssal sedimentation rates.

The observed dramatic increase of the sedimentation rates since Middle Pliocene time could be explained qualitatively by the following sequence of events:

1. By at least the beginning of the Middle Pliocene time the SCB's oceanic fragment dimensions, with edges subducting under the neighboring microplates, reached critical minimum values when, in accord with Hooke's elastic law, the fragment could flex and not just subduct only at its edges. The applied stresses then compelled the fragment to overcome local isostatic balance and so to flex (elastically?) as whole.

2. By the beginning of Middle Pliocene time, a few separated shelf (slope) zones of the Circum-SCB began to unite into one circular crater-shaped shelf zone, yielding a circular shelf cone, with slopes directed to the SCB's center from all sides. In some parts (e.g. to the west) the slope kernels were created by an accretionary melange complex of subduction zones. This growing circular shelf cone constituted a considerable loading of the sediment cover on the basement. Non-abyssal sediment cover was loaded onto the abyssal oceanic-crust basement. This rapid loading was able to act as an additional contribution to removal of local isostatic balance.

The joint development of the two effects in superposition caused a cumulative trigger effect, provoking the dramatic increase in basement subsidence, the accompaniment of turbidite type of sedimentation, and the sedimentation rate increases by almost one order of magnitude.

4. Mud volcanoes and diapirs

As described in the previous section, the South Caspian Basin formed and developed under very interesting geological conditions and, as a result, tremendous sedimentation rates (along with the predominantly shale lithology) occurred, causing massive overpressure and very low temperature gradients. (At least in the zone of the Caspian Sea and adjacent area such abnormal geological conditions led to the formation of a deep (8-9 km) laterally extended zone of decompaction and diapirism).

A decompaction zone, making the system highly unstable, developed in the shaly formations, mostly Miocene-Paleogene (Guliyev, 1997). The mechanism of such formations is not clear. An explanation due to the generation and accumulation gas is not consistent because natural gas at such depths is mostly in solution and its role in influencing the bulk density of rocks is negligible.

Most likely such a zone of decompaction is formed because of high rates of sedimentation. Different types of shale have different abilities to lose water under changing pressure conditions which could lead to the expulsion of water from one formation to others.

Similar phenomena, but in the vertical direction, can be observed in and around mud diapirs (Bagirov and Lerche, 1997a). Typical of modern accretionary wedges, such effects play significant roles in structural development, mass motion, heat transportation, fluid pressure development and hydrocarbon generation, migration and accumulation in the South Caspian Basin.

The mud diapirs are very noticeable on seismic cross-sections. Being kilometers in diameter, they have depths of tens of kilometers. The Abikh and Vezirov diapirs, situated in the central part of the South Caspian Basin on a 12-

second seismic line, can be traced as huge bodies of 10 km width and crossing all 12-seconds of the section, corresponding to a depth of 26-26 km (Fig. 1.4). During their development such structures replace and deform rock formations, which they penetrate. As a result of such strain, large deformations are caused in sedimentary formations, which can lead to failure of the rocks and shallow earthquakes. Such shallow earthquakes are fairly common in the South Caspian Basin, as will be shown later in this volume.

Diapirs are mud bodies, unconsolidated masses of high porosity and, consequently, of relatively high permeability. An excess pressure in such "open" structures cannot be too high and, at the least, is lower than in surrounding formations. Such instability leads to the situation where fluids from surrounding sedimentary formations flow to the body of the diapir. Numerical models show that the flow rates can reach the order of magnitude of meters per year. Naturally, the first to arrive at a diapir is gas, because of its low viscosity and high buoyancy. Then water arrives followed by oil (Bagirov, Nadirov and Lerche, 1998). As a result the body of the diapir is absolutely saturated by gas. Water flowing to the diapir merges with mud rocks and increases the volume of the diapir. The diapir grows and develops this way. The oil mostly remains at the flanks of the diapir and does not enter the diapir. The consequent thermal anomaly caused by the low thermal conductivity of the diapir then changes the hydrocarbon generation picture as well.

Thus, thermal and pressure anomalies change the generation and migration regime of the zone around a diapir in such a way that the diapir turns itself into a giant gas accumulator, while oil accumulations are situated on the flanks of the diapir. Now gas can not fill the structure indefinitely. Eventually gas will find a vent to escape. And such vents are called mud volcanoes. Those volcanoes are the shallow manifestation of the diapirs. There are hundreds of such volcanoes in the South Caspian Basin. According to Dadashev (1995) there are about 260 recorded volcanoes in the territory of Azerbaijan alone. That number is about a third of all mud volcanoes known in all basins of the world. The volcanoes are manifested as hills and raised areas, sometimes reaching upwards of 400 m in height.

Volcanoes lying offshore differ morphologically from those on land, forming islands (7 such islands currently occur in the Caspian Sea) or submarine banks. At the time of an eruption some volcanoes form new islands, which are eroded within several days.

Mud volcano activity can be divided into two stages: gryphon-salse and eruptive. The first stage involves a gentle emission of relatively small quantities of water, mud and gas from gryphons, domes and salses lying on the mud volcano. Some gryphons emit oil, or an oily film, onto the surface of the water.

In contrast to the gryphon activity where the driving force of the mud and fluids is mostly overpressure, during eruptions a huge mass of mud and rocks is pushed by the energy of gas going to free phase. This effect is similar to a champagne bottle, where the overpressure is not significant, while gas bubbles can push the wine until the bottle is almost empty. An eruption also can have a different character. Some eruptions are associated with ignited gas with flames of hundreds of meters in height. In other cases only mud is

13

flowing. Those mud flows can be several meters thick and are spread over hundreds of meters (occasionally several kilometers) in length.

Mud can be semi-dry and its "eruption", or squeezing out, can last years. Sometimes the crest of the crater is lifted for tens of meters before an eruption. And the crater usually subsides after an eruption; this occurs because of the evacuated space formed in the crest of the diapir during an eruption. This combined process causes fracture formation which can be up to a meter or more wide, with offsets of several meters, and lengths of several kilometers.

Offshore eruptions have a similar structure. Crests of the diapirs before eruption can rise so that they are higher than water level and so form islands, which then disappear later. Sometimes the islands can be formed by ejected mud. Gas hydrates are another phenomenon related to volcanoes. The low thermal conductivity and saturated gas cause very low thermal gradient in the diapirs. Together with high pressure and high gas content, this situation provides a very favorable situation for hydrate formation. During dissociation of hydrates a significant amount of energy is released rapidly, enough to make a "hole" in the sea through which to eject gas and sediments, and to heat the gas until ignition temperature is reached causing flame eruptions. Whether hydrates exist in onshore diapirs is not clear and only direct drilling can answer this question.

Here we illustrate the geological situation in the South Caspian Basin, which creates a wide variety of the aspects of geological hazards. The goal of the book is to estimate quantitatively the scales of hazards. Each chapter of the book is devoted dominantly to one or more of the hazard aspects, but most have a mutual interrelationship.

2 EARTHQUAKES HAZARDS

Earthquakes hazards is the problem of highest priority when planning safety operations. All companies take into consideration the possibility of earthquakes when designing their platforms, pipelines and other technical infrastructure. The questions arising here are: How strong an earthquake can occur in a specific area, and how often is it expected? In other words, what is the waiting time for such a strong event? To investigate that problem statistical analysis has been used. Not just the strength distribution, but also the directions of longitudinal waves coming to specific fields, were considered together.

Based on data since 1832 from 533 earthquakes in the Azerbaijan region, in this chapter an analysis is given of: (a) the occurrence likelihood of weak, medium and strong earthquakes, the latter capable of causing significant damage; and (b) the likely directions from which damaging earthquake waves can arrive.

The Chirag region of the South Caspian Basin is used to illustrate application of the methods because of the potential significance of the region for hydrocarbon exploration involving rigs susceptible to the above hazards.

The statistical information would indicate that the occurrence likelihood for a 7-balls or higher damaging earthquake in the Chirag region corresponds to an average waiting time of around 5,000 yrs; a medium strength (6-balls or higher) earthquake should occur, on average, in the region every 1,200 yrs, while a weak earthquake (5-balls or higher) is likely to occur, on average, every 110 years.

The most likely direction of longitudinal seismic waves from earthquakes of sufficient strength to cause significant damage in the Chirag region, be the earthquake epicenter in the region or at a remote focus, is roughly east and west, with a slight prevalence for a westward origin.

In view of the potential hazards for oil rigs in the offshore South Caspian Basin, it would seem that organized, high quality, data collection for both offshore and onshore earthquakes should be rapidly undertaken as a vital adjunct to drilling operations in order to sharpen the assessments of risk factors presented here.

Elsewhere in this volume an investigation is given of the risk of encountering activity in the form of mud flows, breccia ejection, or flame ignition of released gas in the South Caspian onshore and offshore regions of Azerbaijan, based on statistical treatments of historical data from about 220 mud volcanoes. That study was undertaken as a guide to geological hazard assessment for rig siting in the Caspian area (see also Bagirov et al., 1996).

Earthquake hazards which are evaluated here for likelihood of occurrence at a scale sufficient to cause rig damage are: (i) earthquakes on their own; (ii) interrelated earthquake and mud volcano hazards. As will be referenced later, detailed catalogs, atlases and bulletins are available documenting the historical record of these hazard types and are used extensively throughout this chapter.

The offshore Chirag region of the Azerbaijan section of the South Caspian area is under active consideration by oil companies as having high hydrocarbon potential, and will be used to exemplify application of the hazard methods developed here to a potential exploration/exploitation arena.

A. Seismicity Hazards

Apart from the hazard of mud-volcano eruptions in the Chirag area, earthquakes also pose a threat because the Chirag area lies on a seismically active zone formed by the Apsheron-Cheleken sill, an offshore continuation to the east-southeast of the Greater Caucasus. Tectonic and seismic activity are caused by subduction of the South Caspian paleo-oceanic microplate (Nadirov et al., 1997), characterized by a high level of seismic activity and by major earthquakes. Probability distributions of seismic factors also require an evaluation of how representative are the data used. Four seismicity attributes are investigated here:

  1. Frequency and strength distributions of major earthquakes;
  2. Directional distribution of longitudinal waves of major earthquakes;
  3. Possible seismic dislocations and their likely directions;
  4. Evaluation of the seismicity hazard to offshore installations.

1. Frequency and strength distributions of major earthquakes

Earthquakes of varying intensity have been recorded around the Apsheron-Cheleken sill for over 150 years, and have epicenters which are both local and at distant sites, lying within the Elburs, Greater Caucasus and Kopet-dag fold-belts. The seismicity has been analyzed in terms of energetic classes (K) and magnitudes (M), using compiled observations from a network of seismic stations, published in various monographs, atlases and bulletins, variously compiled by the Moscow publishing firm, Nauka, under "Earthquakes in the USSR in 19xx" where xx varies yearly from 77 through 90. The same firm has put together three compilations: "Atlas of Earthquakes in the USSR (1911-1957)" published in 1962, "New Catalog of Major Earthquakes in the USSR from ancient times until 1975" published in 1978, and "Seismic Regions of the USSR" published in 1968.

From this information earthquakes records have been compiled covering the period from 1832-1990, during which 533 earthquakes of varying energetic classes occurred within an area of approximately 10^5 km^2 around the Chirag area.

The energetic class distribution of earthquakes (Table 2.1) shows that a relatively small number of high-energy earthquakes was observed against a general background of weak seismic events. No unambiguous relationship is apparent between the numbers of high-energy and low-energy tremors.

Table 2.1. Earthquake Intensity Distribution at the Epicenter

J (balls)	0	1	2	3	4	5	6	7	8
Total earthquakes	1	16	143	180	70	66	44	10	3
Since 1975:	1	15	136	146	34	18	2	2	0

Representative energetic classes are time-dependent due to recording coverage. In 1831 only earthquakes with K > 12 are noted; by 1951 measurements to K=11 could be made; by 1961 measurements to K=10, and from 1975 onwards measurements to K=9. Early earthquakes of lower energetic classes may not have been recorded due to the sparse coverage by seismic network stations and the low sensitivity of equipment at early times. In the Chirag area there is a high probability that the data analyzed will be incomplete. Statistical analysis methods are then used to assess the omissions.

A map of earthquake epicenters in the Chirag region is given in figure 2.1, adapted from Panakhi and Kasparov (1988). The map was compiled from a catalog of earthquakes covering the period 1832-1982. Earthquakes from this catalog, occurring in the region between 39-42°N and 48-52°E, were culled for a more detailed study.

Each earthquake focus was calculated, and the probability distribution of the level of influence in the Chirag area studied. The calculations were undertaken using an equation (Shebebalin, 1968) describing the intensity, J, for a macroseismic field:

$$J = bM - v \lg (\Delta^2 + h^2)^{1/2} + c \qquad (2.1)$$

where h is the depth of the focus, Δ is the epicenter distance, M the magnitude, and the values of the coefficients, b, v and c differ for different regions. Kuliev (1978) has recommended b=1.5, v=3.3, and c=2.7 as good statistical values for Azerbaijan for all periods of observation.

Earthquake data from after 1975, when a larger number of weaker earthquakes (K=9) became representative, have been used to construct the intensity both at the epicenter (Δ=0) of each earthquake and at the location of the Chirag area. The results are displayed in Figs. 2.2 and 2.3, and in Tables 2.1 and 2.2, respectively.

The maximum intensity reached 8 balls at the epicenters of 3 earthquakes occurring from 1832-1990 in the adjacent Caspian Sea. The intensity level of the same earthquakes reached only 5 balls in the Chirag area.

18

On the basis of the data available a statistical prediction is now given of the susceptibility of the Chirag area to major earthquakes.

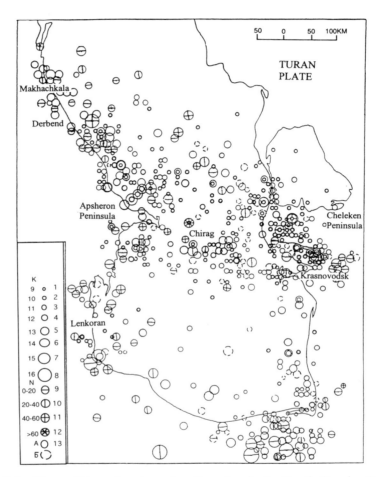

Figure 2.1. Chart of the epicenters of earthquakes in the South Caspian Basin.

Table 2.2. Earthquake Intensity Distribution in the Chirag Area

J (balls)	0	1	2	3	4	5	6	7
Total earthquakes	258	121	86	53	14	1	0	0
Since 1975:	225	79	34	13	3	0	0	0

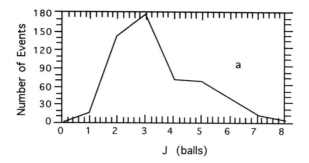

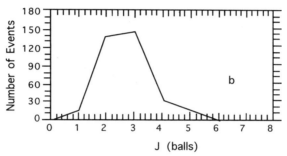

Figure 2.2. Distribution of earthquake intensity at the epicenters: a) for all periods of observation; b) after 1975.

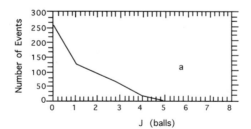

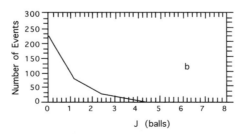

Figure 2.3. Distribution of earthquake intensity in Chirag: a) for all periods of observation; b) after 1975.

Denote by A_1 an event of ≥ 8 balls, by A_2 an event of ≥ 7 balls; by A_3 an event of ≥ 6 balls and by A_4 an event of ≥ 5 balls. These event classes are cumulative in the sense that an A_1 event automatically includes all lower ball events.

Let p_i be the probability of an A_i event during some small time interval. Here the interval is taken to be 1 second, small compared to earthquake lifetimes, which will improve the approximation. Then the probability that an A_i event will occur k times in a year is:

$$c_n{}^k p_i{}^k (1-p_i)^{n-k} \qquad (2.2)$$

where n is the number of seconds in a year ($\sim\pi.10^7$), and $c_n{}^k = n!/[k!(n-k)!]$. Because n is large, and p is small, the probability (from equation (2.2)) of an A_i event in a time of t years is well described by the Poisson process:

$$\exp(-\lambda_i t)\,(\lambda_i t)^k/k! \qquad (2.3)$$

where $\lambda_i = p_i n$, is the expected number of events per year.

Table 2.2 indicates that A_1, A_2, and A_3 events have not been recorded in the Chirag area over the past 158 years (1832 through 1990), but the probability of these events is not zero. The probability of earthquake epicenters with corresponding intensities developing in the Chirag area must be evaluated.

From Table 2.1, A_1 events occurred 3 times over 158 years in a total area $A_T = 113,220$ km^2, A_2 events 13 times; and A_3 events 57 times. Then, for an area $A_c = 270$ km^2 (the Chirag area), λ_i may be evaluated as:

$$\lambda_1 = (3A_c)/(158A_T) = 4.5 \times 10^{-5}\ \mathrm{yr}^{-1};$$
$$\lambda_2 = (13A_c)/(158A_T) = 19.6 \times 10^{-5}\ \mathrm{yr}^{-1};$$
$$\lambda_3 = (57A_c)/(158A_T) = 86.0 \times 10^{-5}\ \mathrm{yr}^{-1}.$$

Determination of λ_4 (associated with A_4 events) is more complex. First, over the last 158 years, an A_4 event did occur in the Chirag area; and second, it is necessary to take account of the representativeness of the early data for 5-ball earthquakes. Data from the past 15 years, over which period 18 earthquakes of 5 balls have occurred in the total area, allow one to write

$$\lambda_4 = 1/158 + (18A_c)/(15A_T) = 9.19 \times 10^{-3}\ \mathrm{yr}^{-1}.$$

Figure 2.4 shows the expected number of events of classes A_1, A_2, A_3 and A_4 over t years; while figures 2.5(a-f) show the probability that, over t years, no 8-ball earthquakes or above will occur; no 7-ball or above; no 6-ball or above; and no 5-ball or above, respectively. Figures 2.6(a-d) show the waiting time distributions until the first occurrence of events of classes A_1, A_2, A_3 and A_4. Note that the average waiting time for each class is just λ_i^{-1}, which is 22,000 yrs for an A_1 event, 5,000 yrs for an A_2 event, 1,200 years for an A_3 event; but just 110 yrs for an A_4 event.

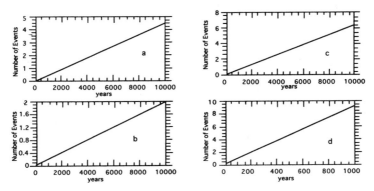

Figure 2.4. Expected number of earthquakes: a) ≥ 8 balls; b) ≥ 7 balls; c) ≥ 6 balls; d) ≥ 5 balls.

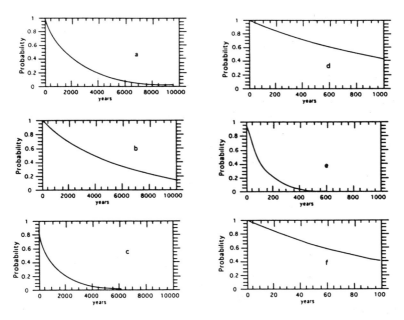

Figure 2.5. Probability of non-occurrence of earthquakes: a) $J \geq 8$; b) $J \geq 7$; c) and d) $J \geq 6$; e) and f) $J \geq 5$.

22

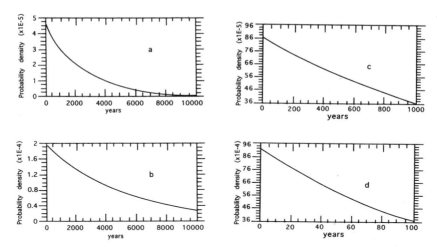

Figure 2.6. Waiting time distribution until the first earthquake occurrence: a) J ≥ 8; b) J ≥ 7; c) J ≥ 6; d) J ≥ 5.

2. Directional Distribution of Longitudinal Earthquake Waves

The seismicity of the Apsheron-Cheleken sill was massively detailed by Agamirzoev (1976), and Agamirzoev and Gyul' (1972, 1973a,b, 1983), while the Apsheron-Cheleken seismic zone was traced by Panakhi (1988). Two seismic zones, northern and southern, were identified with the northern seismic zone extending in an approximately east-west direction along the northern margin of the Apsheron Peninsula crossing the Caspian Sea, and continuing along the northern margin of the Cheleken Peninsula. The zone is caused by a crustal fault between two blocks, and is regarded as a continuation of the Main Caucasus Fault. Transverse faults separate the zone into four seismic structures. The southern seismic zone extends eastward, in an approximately east-west direction, from the southern edge of the Apsheron Peninsula to the southern coast of the Cheleken Peninsula, and is parallel to the northern zone. The zone is an easterly continuation of the Vandam disjunctive fault (Nadirov et al., 1996). Transverse faults separate the zone into four seismic structures (Fig. 2.7). Local foci earthquakes result from movements along inter-block faults. The seismicity of the entire zone results from both local and remote foci.

Panakhi and Kasparov (1988), Agamirzoev (1976), Panakhi (1988), and Agamirzoev and Gyul' (1972, 1973a,b, 1983) have shown that seismogenic structures within this zone are capable of generating earthquakes of up to 8-9 balls. However, distant seismogenic structures, characterized by frequent, powerful earthquakes (9-10 balls), may also pose a seismic threat to offshore installations.

Earthquakes from remote foci originate from zones beyond the Apsheron-Cheleken sill. Examples are the Krasnovodsk earthquake (M=7-8;

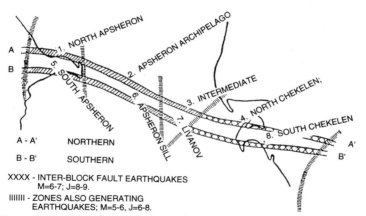

A - A'   NORTHERN

B - B'   SOUTHERN

XXXX - INTER-BLOCK FAULT EARTHQUAKES
      M=6-7; J=8-9.
IIIIIII - ZONES ALSO GENERATING
      EARTHQUAKES; M=5-6, J=6-8.

Figure 2.7. Seismogenic zones of the Apsheron-Cheleken sill.

J=9-10 balls), generated within the Makhachkala-Krasnovodsk seismic zone; the Agrakhan-Derbend, Lekoran-Gorgan and North Elburs seismic zones with potential seismicity of 8-9 balls. Their influence on the Chirag area may be marked.

The directional distribution of longitudinal earthquake waves in the Chirag area was calculated using both the values of the seismic intensity of all the earthquakes on the Chirag area, and the azimuthal directions. The results were subject to statistical analysis in order to determine the different azimuth directions. Graphical results used an azimuthal rose diagram with angular intervals of $\pi/8$. The frequency of occurrence, N_i, in each angular interval was calculated from

$$N_i = \sum_{i\alpha-\beta < x_j \leq i\alpha+\beta} j, \qquad i=1,2,.....16 \qquad ; \qquad (2.4)$$

where x_j is the azimuthal direction from which the longitudinal wave earthquake reaches the Chirag area at latitude and longitude coordinates $\alpha=22°15'$, $\beta=11°12'30"$, respectively; and where j is a counter for earthquake numbering.

Table 2.3a presents the statistical data with which the azimuthal rose was constructed, while figure 2.8 presents the azimuthal rose, reflecting the distribution of the angles from which longitudinal waves are most likely to arrive. It is apparent from figure 2.8 that the greatest number of earthquakes have arrived from the northwest and west; i.e. from the direction of the Greater Caucasus. However, figure 2.8 does not reflect directions of the most hazardous longitudinal waves, i.e. because the majority of earthquakes are weak or because of their large distances from the Chirag area they may not be felt. Therefore, a second rose diagram was constructed (Fig. 2.9) using the data from Table 2.3b, in which events are weighted by intensity, J_j. Then one can write

$$N_i = \sum_{i\alpha-\beta<x_j\leq i\alpha+\beta} J_j \ ; \quad i=1,2,\ldots.16 \quad . \tag{2.5}$$

Here, J_j is the effect (in balls) of the j-th earthquake occurring in the i-th angular interval.

Table 2.3. Earthquake Directions and Numbers

	By Number	By Strength
Direction	a	b
N	12	10
NNW	34	45
NW	64	162
WNW	84	88
W	81	92
WSW	33	37
SW	17	12
SSW	2	2
S	1	1
SSE	1	1
SE	9	3
ESE	8	4
E	27	42
ENE	37	16
NE	24	11
NNE	9	7

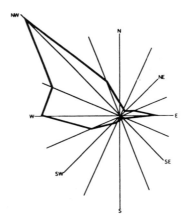

Figure 2.8. Azimuthal rose diagram of the number of earthquakes felt in the Chirag area.

It is apparent from figure 2.9 that the most powerful shocks from longitudinal waves are expected from the east and west directions.

These results help in choosing the optimum directions for offshore construction in order to minimize the effects of probable, powerful, destructive earthquakes.

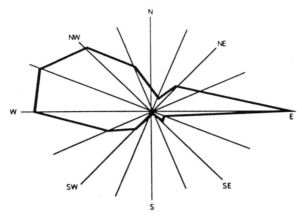

Figure 2.9. Azimuthal rose diagram of the angular distribution of the product of the intensity and number of earthquakes felt in the Chirag area.

3. Occurrences of Seismic Dislocations, Causes, and Spatial Orientations

Earthquakes result from elastic deformation occurring in areas where a rock mass does not yield to plastic deformation or compaction, causing sliding-deformation and thereby generation of elastic seismic waves. The determinations of the mechanisms at the foci of 17 earthquakes, worked out by Agalarova (1969), are presented below. The low number of determinations is because there is an absence of macroseismic data on earthquakes in the Caspian Sea, and only a few foci located on land have such data. The absence of a dense seismic network, weak and indistinct records of first arrivals, and the absence of data from deep foci, also influence the ability to provide more than 17 investigations.

The position of the main stress axis active at the focus, and also the two possible positions of the fault plane, were determined for each earthquake on the basis of the observed distribution of displacement indicators. The nodal lines of P-waves represent two mutually perpendicular arcs of a great circle, corresponding to lines cutting two mutually perpendicular planes of a sphere. The results are presented in Fig. 2.10. Features of the stresses at the foci of two of the earthquakes are now considered.

(i) Caspian earthquakes occurred in 1961 and 1963 from the same focal zone, with coincident epicentral coordinates. One of the possible fault planes dips steeply (60°) to the southwest; movement along this plane would have a normal-wrench character, with a predominance of normal fault movement. The second plane has a nearly north-south orientation and dips to the southeast at 40°; movement on this plane has a normal-wrench character, and the hanging-wall would subside. Movements along both of the possible

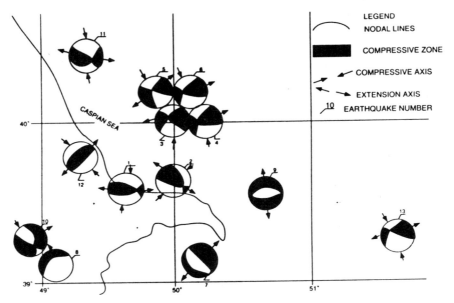

Figure 2.10. Mechanisms at earthquake foci.

fault planes are associated with subsidence of the hanging-wall of the fault and uplift of the footwall.

(ii) Earthquake of 6 March 1986 (h=48 km). This earthquake occurred beneath the Caspian Sea and was accompanied by a large number of secondary tremors, of which 65 had K ≥ 10, and 25 of the 65 had K ≥ 11. No similar seismic activity has been observed under the Caspian Sea during the past 100 years, although a major earthquake (M=6.2) was recorded on 27 January 1963, and occurred with a maximum intensity of 7-8 balls, in the Neft Dashlary area. The focal zone of the earthquake lay within the deepest part of the Central Caspian (water depth about 200 m), and is associated with the junction between the zone of alpine folding in the Kopetdag-Caucasus area and the epihercynian Turan plate.

Investigation of earthquakes on the Turkmenistan side of the Caspian Sea has shown that the primary seismic dislocations reach the surface as dextral shears with small displacements (5-30 cm). En-echelon anticlines on the Apsheron-Cheleken sill, in the interval from Cheleken to Bezymyannaya Bank, indicate that dextral shear also characterizes the eastern part of the Apsheron-Cheleken fault. These deformations are only short-lived manifestations of an extended process of dextral shifts, the total magnitude amounting to about 1 km over Holocene time.

The epicenter of the 6 March 1986 earthquake indicates that the event is associated with a point where the northern zone transects the Central Caspian Fault. Some of the aftershocks occur below the Moho, demonstrating that the Apsheron-Pribalkhan subduction zone has deep roots, but the majority are concentrated within the crystalline part of the plate and in Mesozoic rocks, with relatively few in the Cenozoic molasse; most have a west-northwest orientation.

The structural characteristics of the majority of the foci of Caspian earthquakes in this zone indicate that the dominant earthquake cause is active tectonic movement along the boundary between two major crustal structures, the Turan plate and the South Caspian paleo-oceanic microplate.

Accordingly, there is a possibility of seismic dislocations affecting the Chirag area. However, the absence of sufficient information on the earthquake mechanisms in the immediate vicinity of Chirag prevents a determination of the prevailing local crustal stress fields. At the moment only a cautionary note is possible, warning of the potential for seismic dislocations.

4. An Assessment Procedure for Seismic Hazard to Installation Foundations

An evaluation of the seismic hazard to installation foundations during well drilling, allowing for acoustic rigidity, can be made (in principle) with sufficient data. Here the logic procedure is provided to indicate what data are needed to carry through the assessment. The initial seismic intensity for a specific site area is predicted on the basis of seismological and seismotectonic studies, taking into account the mechanical properties of the sediments and the standing level of overlying water. An important point in calculating the absolute seismicity (in balls) for each site is to establish a "standard sediment" response.

Evaluation of the absolute seismicity, s (in balls), at each site is carried out with an acoustic rigidity method using Medvedev's formula:

$$s = 1.67g(v_{st}\rho_{st}/v_i\rho_i) + \exp\{-0.04h^2\} \qquad (2.6)$$

where v_{st} and ρ_{st} are the spreading velocity of longitudinal waves and the density of the standard sediment, respectively; v_i and ρ_i are the P-wave velocity and density for the non-water-saturated sediments, respectively, and g is the acceleration due to gravity. In the case where a water layer is present, an extra intensity is calculated from the formula:

$$\Delta s = \exp\{-0.04h^2\}$$

where h is the water level (in m); where there is total water saturation (as on the sea bed), h=0 and Δs=1; i.e. one ball is added. For multi-layered sediments, of total thickness H, the equivalent acoustic rigidity is calculated from the formula:

$$v_i \rho_i = \sum_n (v_n \rho_n h_n)/H \qquad\qquad (2.7)$$

where h_n is the thickness of each layer; v_n and ρ_n are the velocity and density, respectively, of each layer.

Because there is a layer of muddy deposits on the sea bed down to a depth of about 5-7 m, the calculations need to be done for foundation-supporting depth intervals below this layer. When sediment samples from wells become available, and also seismic measurements of acoustic rigidity, the above method can be used to assess the seismicity.

Care must be taken to be aware of the effects of prior mud volcanoes which can influence seismic sensitivity. Periodic eruptions of breccia and mud flows from mud volcanoes over Quaternary time can lead to the formation of argillaceous layers reaching several hundred meters in size (Bagirov et al., 1996). Such layers can act as acoustic shadows, altering the surrounding seismic sensitivity. Detailed consideration of this problem is beyond the scope of the present volume.

B. Conclusions

The Chirag area of the South Caspian Basin was taken as an example to demonstrate the seismic hazards assessment, lies in a seismically active zone, and earthquakes with intensities of up to 8 balls could occur. An analysis of earthquakes since 1832 has shown that the probability of such earthquakes is low; the probable recurrence time of greater than an 8-ball earthquake is 22,000 years; for 7 to 8-ball earthquakes, the corresponding period is 5000 years; for 6-ball and above, 1200 years; and 5-ball and above earthquakes have a probable recurrence time of about 110 years.

The directions of longitudinal seismic waves indicate that the most probable directions from which damaging shocks can reach the Chirag area is east and west.

In producing the probabilistic method for prediction of earthquakes, no account was taken of gaps in the data, which would be an interesting task to investigate with a better data base. It would also then be of use to construct a map of the seismo-tectonic regions around the Apsheron-Cheleken sill and adjacent areas, allowing for new data and remote, seismically active, structures. Such a map would provide the level of seismic hazard in the Caspian Sea and in the adjacent regions of Azerbaijan and Turkmenistan, and would be useful for calculating the seismic hazard to offshore and coastal installations.

In the event that an earthquake (even a weak one) occurs in the Chirag area, it will then be possible to calculate the mechanisms at the focus of the earthquake, to determine the compressive and extensional stresses in the Chirag area, and to calculate the probability of forming seismic dislocations.

When sediment samples become available from wells in the Chirag area, it will be possible to measure the values of acoustic rigidity and so to evaluate the absolute seismicity. The effect of plastic argillaceous bodies, formed by mud volcanoes, on the total acoustic rigidity of the medium is also an outstanding area of concern requiring more data for its resolution. Such results would be of benefit in planning and siting platforms and other offshore installations.

Perhaps this work, and that of Bagirov et al. (1996), will have more than served their purposes if good, high quality, seismic activity data are routinely collected in the future to sharpen the points which are often but dimly seen with the current data base. In view of the hazard potential for oil rigs in the offshore Caspian Sea, it is likely that such data collection should be rapidly undertaken as a vital adjunct to drilling operations.

3 Fault and Fracture Hazards

The hazards discussed here are associated mostly with the sensitivity of mud volcanic activity, while the range of geological and tectonical reasons cause the faulting and fracturing is much wider. As was shown in Chapter 1, the unique geological conditions of the South Caspian Basin led to a complex situation, where rising diapirs have an impact on the surrounding sedimentary formations, causing large strain and stress. If the stress factor is strong enough for rock failure, it will cause some of the fracturing and faulting associated with shallow earthquakes. On the other hand, earthquakes alone can cause an eruption of a mud volcano. A small dislocation at the volcano vent will be like opening a "bottle cork", so if the diapir is charged, then an eruption will start.

Eruptions have a large influence on the surface. Erupted mud shows large stresses on the surface deposits around a crater. Then after the eruption, when thousands or millions tons of mud have been brought to the surface, the top of the volcano can sink or subside, and that causes more surface fractures. Those fractures, extending sometimes to several kilometers in length, displace the ground for several meters, presenting a hazard to technical infrastructures.

Based on data from more than 200 mud volcanoes in the Azerbaijan region, an analysis is given of: (a) the likelihood of a mud volcano hazard (ejected breccia and/or mud flows and/or flame ignition) in temporal association with an earthquake; and (b) the likelihood of fracture formation associated with mud volcanic eruptions.

There is some correlation between earthquake actively and mud volcano activity, suggesting that mud volcanoes occur between zero to five years prior to earthquakes. But the correlation is not sharply delineated due to the paucity and quality of currently available data.

For surface fractures, associated with mud volcano eruptions, which can be meters wide and can stretch for a kilometer or more, only volcanoes with five or more eruptions were used to estimate the likelihood of fracture occurrence, yielding an average of about 30% chance of occurrence. The low amount and low quality of data did not permit any more detailed investigation of fracture parameters - such as average width, length or offset.

In view of the potential hazards for oil rigs in the offshore South Caspian Basin, it would seem that organized, high quality, data collection both offshore and onshore should be rapidly undertaken as a vital adjunct to drilling operations in order to sharpen the assessments of risk factors presented here.

In addition to the earthquake hazards discussed in Chapter 2, two further hazards are: (i) interrelated seismic and mud volcano hazards; and (ii)

31

fracture and fault formation caused by mud volcanism. The bulletins available, documenting the historical record of these hazard types, are those used extensively in Chapter 2.

A. Deformation of Sedimentary Beds Around a Mud Diapir

The evolution of mud diapir structures has a significant impact on the overlying and surrounding bed formations. The inverse quantitative procedure described in Lerche et al. (1997) allows for estimation of the interactive evolution of mud and sediments through time. The number of time-steps is determined by the number of observed beds. The larger the number of beds, the larger the number of time-steps required and the closer the changing shape of the mud surface can be modeled in time. Accordingly, the deformation of the surrounding sediments can be followed with the same accuracy. This capability provides the option of inferring and following the strain in the formations with time, unlike the majority of mathematical models that use the estimated stress (based on assumptions of the behavior of the dynamical system) to infer the resulting deformation in the sediments. The deformation of rocks, whether the deformation is permanent or not, and the mode of deformation (for instance folding and/or faulting) depend on the interaction between a number of physical and chemical factors such as fluid pressure, rates of deformation processes, rock composition, cementation, temperature, etc. The stresses that generated the deformations in the formations can themselves then be inferred through application of the theory of elasticity; the state of stress during time of diapir rise can thus be estimated in this way. This approach was presented by O'Brien and Lerche (1987) where draping of sediments around a simplified salt shape was examined.

When a block of sediment is exposed to stress due to the influence of the load of the overlying sediments and to the impact of the rising mud diapir, the individual particles are displaced to new positions. The progressive deformation of sediments in the vicinity of a rising diapir structure can be any combination of an overall translation of a sediment unit, together with local distortions and rotations (Hobbs et al., 1976). The deformation in a sediment unit (as a response to the stress field) will continue until an equilibrium in the configuration of the sediment units has been reached. However, once an equilibrium is reached a state of stress may still remain. In order to track the deformation paths of the sediments, the beds are subdivided into smaller units. The degree to which the beds can be subdivided is limited by the resolution in the numerical system. By following the displacement of the corners describing the sediment volumes from time-step to time-step, the changing displacement field (deformation) can be mapped. The tracking of deformation is thus a study of the change in geometries, i.e. the relative configuration of the corner points. The displacements of the individual particles can be traced from the undeformed state to any deformed state, i.e. the differential strain as well as the accumulated strain can be calculated. This calculation allows for mapping the deformation history of the sediment particles. If the deformed state is compared to the undeformed state the concern is only with the accumulated deformation.

The method of the stress/strain calculations is described in Appendix A. The method can be used both for rising mud and salt diapirs. Here an application to the Abikh diapir will be shown, which is in the central deep-water part of the Basin. A 12-second two-way seismic line crosses this structure from east to west. The diapir penetrates all sedimentary Meso-Cenozoic formations present in the section. The roots of the diapir extend to a depth of 12-seconds on the seismic line, corresponding to the depths 22-26 km, and the crest of the diapir almost reaches the sediment surface; the width of the diapir is about 10 km.

The evolution of strain in sediments neighboring the Abikh diapir can be drawn immediately based upon the geometrical rise of the diapir. The present patterns of differential strain are shown in figure 3.1a. The high values of the strain at present-day are observed in the shallow part of the section near the Abikh diapir.

For the evolving differential stress patterns, associated with the strain evolution, elastic constants connecting strain to stress must be provided. We have followed the prescription of O'Brien and Lerche (1987), and so used the seismic velocities to obtain a broad trend behavior for the equivalent Lamé constants. In addition, the scale-depth over which the Lamé constants progressively increase from their near-surface low values (equivalent to highly unconsolidated sediments) to deep-sediment high values (equivalent to highly compacted and/or highly consolidated material) was set at 9 km. With the P-wave and S-wave seismic velocities, V_p and V_s, related to the Lamé constants, λ and μ, and to local bulk sediment density, ρ, through $\rho V_p^2 = \lambda + 2\mu$; and $\rho V_s^2 = \mu$, the low values for the two Lamé constants, λ and μ, were then set at $\lambda_{min} = \mu_{min} = 2.10^8$ kgm^{-2}s^{-2}, while the fully consolidated values were taken to be $\lambda_{max} = \mu_{max} = 2.5.10^{10}$ kgm^{-2}s^{-2}.

The evolution of differential stress could then be traced using the detailed procedure given in Lerche and Petersen (1995). Shown in figure 3.1b are the differential stress patterns corresponding to the strains of figure 3.1a. The dominant differential stress occurs in association with the sub-horizontal strain at diapiric onset, because an isotropic sediment stress-strain relation is used, the principal differential stress directions parallel those of the differential strain. Maximum values of the differential stress reach $2.5.10^{10}$ Pa. To be noted, however, is the presence of a secondary regime of sub-vertical stress at the present-day, within a radius of the diapir and extending from around 5-19 km depth range. This secondary stress regime is indicative of the presence of late fracturing and/or faulting of sediments in the vicinity of the diapir. Two consequences are immediate from the inferred presence of such fractures and/or faults; first, significant fault motion would lead to the presence of earthquakes with epicenters within a domain of about two diapir radii centered on the Abikh diapir, and with originating centers from 5-20 km depth within the sedimentary pile. This inference is in agreement with the statistical information gleaned from the historical record alone, as given in earlier chapters. Second, the generation of fractures provides high permeability pathways for the escape of hydrocarbons from depth to the shallower horizons of the petroliferous-bearing Productive Series (0-9 km); and, at shallower depth, for the initiation of fractures leading to escape of hydrocarbons to the surface, where they dissipate.

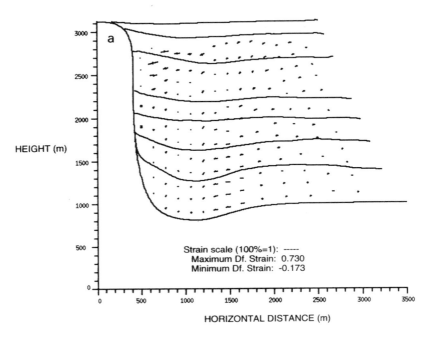

Figure 3.1. a) Differential strain in the sediments around the Abikh diapir.

The accumulated strain and stress are shown in fig. 3.2a-b. In this case the presence of the large sub-horizontal accumulated stress/strain completely dominates the ability to see the stress/strain build-up in later sediments, which are better depicted in the differential form. But the dominance of the stress/strain in the early Middle Pliocene sediments suggests that these sediments should have low mechanical strength, implying a deep zone of relatively low seismic velocity.

The Coulomb-Mohr criterion (Lerche and Petersen, 1995) for rock failure was also applied to the sedimentary evolution of stress and strain. The inherent rock shear strength was set at 3×10^8 kgm^{-2}s^{-2} and the coefficient of internal friction at 0.65, with both values being taken from Jaeger and Cook (1976).

Because the inherent shear strength for rock failure is only slightly above the minimum values for the Lamé elastic constants λ and μ, it might be expected that failure would occur dominantly throughout the sedimentary pile, and so one should not be able to see any significant differences in principal and

secondary strain or principal and secondary stress amounts. Figure 3.3 shows
the possible directions of faulting for present-day.

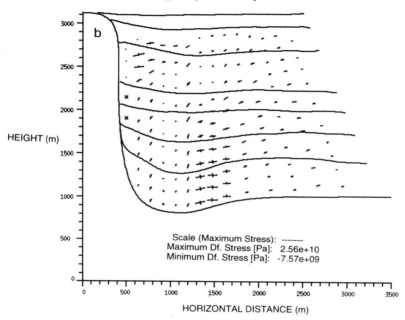

Figure 3.1. b) Differential stress in the sediments around the Abikh diapir.

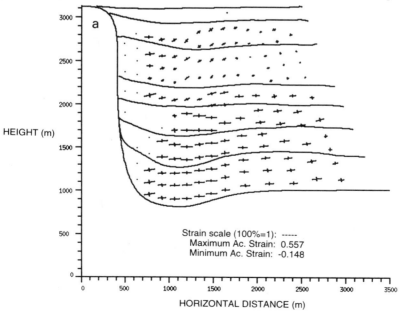

Figure 3.2. a) Accumulated strain in the sediments around the Abikh diapir.

36

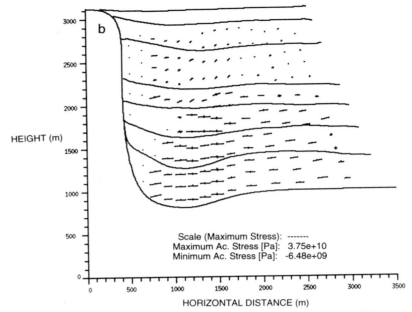

Figure 3.2. b) Accumulated stress in the sediments around the Abikh diapir.

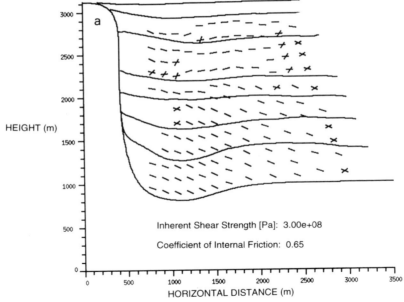

Figure 3.3. Directions of possible failure.

In short: the evolution of stress and strain in the sediments bordering on the mud diapir would indicate a deep zone (pre-Middle Pliocene) of sub-horizontal rock failure, suggesting a deep zone of low seismic velocity caused by sediment infill of the secondary rim syncline produced by the diapiric rise. The production of shallower (5-20 km) regions of subvertical stress and strain in post-Middle Pliocene sediments suggests that sediment-induced earthquake epicenters should occur mainly within a zone of a couple of radii of the mud diapir, with originating centers at 5-20 km depth. The high stress in these neighboring sediments also suggests that fracture formation should lead to enhanced migration pathways for deeply produced hydrocarbons to reach the Productive Series, and for shallower-produced fractures to cause oil and gas migration to the sediment surface, with subsequent dissipation.

B. Interrelated Seismic and Mud Volcano Hazards

A high level of seismic activity and mud volcanism are both manifestations of energy release from depth, but each is the result of different physical-chemical processes. Interrelationships between these geological phenomena have been investigated. The pioneering work of Abikh (1939) invoked seismic shocks as triggers for mud-volcanic activity. Geological Institute of Azerbaijan personnel have also examined possible interrelationships because powerful eruptions of mud volcanoes have been tied to earthquakes, e.g. Livanova Bank which occurred after the Krasnovodsk earthquake (1895); Shikhzairli which erupted after the Shemakha earthquake (1902); Dzhau-Tepe which erupted after the Crimean earthquake (1927); while four mud-volcano islands appeared in the Makran coastal zone of the Arabian Sea after an earthquake on 28 November 1945. The emission of a tongue of flame to a considerable height, and the occurrence of new islands, were observed in the shelf zone of the South China Sea on Hokkaido Island, Japan after a powerful earthquake. The majority of mud volcanoes occur after earthquakes of 6 balls or more. However, mud-volcano activity has been observed also after minor earth tremors: strong gas eruptions occurred on the Bozdag Kobiiski mud volcano following two minor earthquakes in Mashtagy on 8 August 1953 (3 balls) and Gebele on 11 August 1953 (3-4 balls). Eruption of the Keireki mud volcano on 14 April 1968 occurred after the Ismailly earthquakes of 5th and 9th February 1968; and the Kelany mud volcano became active on 12 December 1969 after an earthquake on 4 November 1969 with an epicenter south of Baku.

It is noteworthy that a zone of high seismic activity and a zone of mud volcano activity coincide in space, and also that they are located in regions where intense mountain-building processes are taking place with subduction and plate collision. It is even more remarkable that both processes are similar in their time development: at first there is a slow process of increasing potential energy (pressures or stresses in the crust) which reaches a critical point and is then suddenly released (in comparison with the period of accumulation), and transformed into kinetic energy of an eruption or an earthquake. The critical values of the energies differ with each occurrence, depending on parameters related to the earthquake focus or to the mud volcano (depth, physical parameters of the medium, etc.).

A well-defined procedure exists for measuring earthquake energies (Richter, 1958; Riznichenko, 1960; Shebebalin, 1955). For mud volcano eruptions no precise methodology is available, although Rakhmanov (1987) has made significant developments.

A substantial proportion of mud-volcano eruptions are accompanied by seismic effects. Malinovskii (1938) provided a method for measuring the intensity of seismic shocks accompanying mud volcano eruptions. Using this method the epicentral seismic effect of an eruption of the mud volcano on Sangi-Mugan (Svinoi) Island on 11 April 1932 reached 6 balls; and eruptions of Kumani Bank on 1 May 1927 and 5 November 1928 were each estimated at 4 balls. However, the energy of the seismic shocks accompanying mud volcano eruptions is only a small fraction of the total energy. More energy is expended in transporting breccia, overcoming frictional forces, and generating fractures (see later).

The qualitative physical basis for the mutual influence of each of these phenomena on the other is now considered.

The occurrence of earthquakes, and the development of associated fractures, change the pressure conditions within the root zones of mud volcanoes where the energy accumulates, and serve as a trigger for mud volcano eruptions. Two major problems need addressing:

1. The correlation in time of events for each of the phenomena and, in the event that a definite correlation is found, to determine the direction and relaxation time.

2. Consider the hypothesis that both phenomena are independent; what is the probability of a joint occurrence within a time interval Δt, and how does the joint probability compare with the frequency of observed coincidences? What allowance needs to be made for the distance in space between events?

To address the first problem, compare in time the following four events: a) eruption of a mud volcano; b) a major earthquake (K=12); c) a minor earthquake (K=11); d) all earthquakes. Individual consideration of minor, major, and all earthquakes is related to the different dates for the beginning of representative earthquake records of different classes (see Section B). For each of the four events, graphs of the moving average are presented in Figs. 3.4 and 3.5, respectively. The more accurate measurement of weak earthquakes over the last decade creates an artifact of an avalanche-like growth in the number of minor earthquakes in recent years, and so in the total number of earthquakes (Fig. 3.4b). Accordingly, comparison of the graphs of moving averages with mud-volcano eruptions is made only for major (K=12) earthquakes.

As shown on fig. 3.5a and b, some coincidences are seen. For instance, the relatively "weak plateau" A lies between years 1860-1870 on the graph of major earthquakes, whereas plateau A lies 5 years earlier on the graph of eruptions. Peak B on the graph of major earthquakes falls at year 1933, and peak C at year 1939; whereas on the graph of eruptions, peak B falls at year 1930 and peak C at 1937, for 3-year and 2-year earlier occurrences, respectively. Peak D falls at year 1957 on the graph of major earthquakes and at year 1956 on the graph of eruptions, the eruption peak is just one year earlier; peak E falls at year 1966 on both the graph of major earthquakes and the graph of eruptions; i.e. the phase lag of the earthquake relative to the eruption is less than one year. But peak F lies at year 1976 on the graph of major earthquakes and at 1973 on the graph of eruptions, so the phase of the eruptions is again

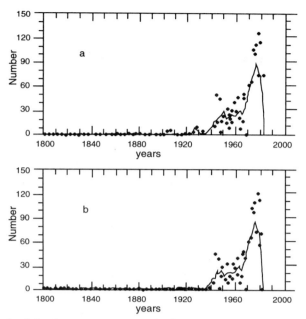

Figure 3.4. a) Moving average graph for the number of earthquakes in Azerbaijan; b) Moving average graph for the number of weak earthquakes in Azerbaijan.

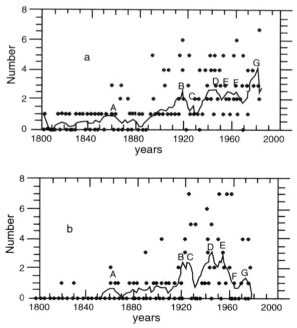

Figure 3.5. Moving average graphs for: a) number of eruptions; b) number of strong earthquakes.

observed to be three years ahead; while peak G coincides on both graphs at year 1986.

From the analyses, the relaxation phase varies from 0 to 5 years; and the peaks of mud volcano eruptions are ahead of those for major earthquakes. However, the analysis to date is not very precise and, while confirming the existence of a mutual relationship between the occurrence of both phenomena, needs to be subjected to a more detailed investigation.

The second problem is more complex. During the period of slow accumulation of energy (stress, pressure) it is possible that, as the moment of energy discharge approaches, earthquakes and eruptions may become dependent, i.e. they could trigger each other. Such situations have been noted repeatedly in Azerbaijan, and the relaxation phase has been several weeks to several months, although no systematic study has yet been carried out.

A frequency distribution for eruptions of different volcanoes is given in Chapter 5, and the frequency of different earthquake classes with time was given in Chapter 2 of the present volume. The next step would be to investigate the joint probability that events from the two types of phenomena fall within a time interval, in order to study the probability of joint occurrence.

The problem is actually even more complex because of the spatial separation between events. The solution of such a complex problem demands a better data base than is currently available.

C. Fracture formations associated with mud volcano eruptions

Mud volcano eruptions are often accompanied by the formation of fractures, which represent a hazard, because their widths may reach of order of meter or more, with displacements of several meters, and lengths of several kilometers. Historical records show how dangerous the eruptions could be from the fracturing point of view. For instance, Agabekov and Bagir-zadeh (1948) described a tremendous eruption of Hamamdag volcano, which happened in September 1947. Fishermen watching that phenomena told that during 5 minutes a frightful underground rumble had been heard, followed by a huge explosion. Along 1 km the coast rose up 5-10 meters. In some places huge shale blocks larger than 20 m^3 fell down to the sea. The area of the fresh mud cover was 400 m x 400 m. Around this cover very deep fractures of 1 m width were formed. A new crater sank 3 m compared to the surrounding area. On the fresh mud cover 17 new salsas emitted water, gas and mud. There were a lot of rock fragments.

Based on the slope of the fresh mud flow, one can define that, at the moment of eruption, all of this mass was lifted to a height of 30-40 meters and then sunk, forming concentric circular subsidence fractures. From the crater to the north at a distance 1200 m (until the sea coast) and 600-800 m in width, there was a strip zone of high fracturing. The fractures had different directions and in some places were so deep that it was impossible visually to define their depths. The edges were lifted or fallen relative to each other and displacements were up to 10 meters or more. Thus, all the area had a very rough topography. This rough zone was separated from the rest (stationary area) by two main faults extending from the volcano. One could see sliding plains on the walls of those faults. In some of the places the horizontal displacement reached 10-15 meters. The vertical displacement increased toward the coast and reached 20 m and more. Looking at the shells, gravel and wave traces one could define that in some places the coast lifted up to 10-20 meters.

A concrete base for a diesel engine was cracked and sloped 30-35° to the horizontal. The pools for drilling mud (having almost 1 meter thick cement walls) also cracked into separate pieces. A 6-inch pipeline was torn apart and displaced in different directions. The road to the rig was destroyed completely.

Some other authors describe similar phenomena on other volcanoes. For example, Yakubov et al. (1965) wrote that during an eruption on Ayrantekyan volcano in 1964 a lot of fractures and faults occurred. The displacement of the largest fault was 2 m with the depth of the open part 0.5-0.7 m. Intensive fracturing also occurred near the old crater field, where there were gryphons and domes. Fractures here destroyed some domes of 1 m height. A number of fractures formed during the eruption of Bolshoi Kanizadag in 1950, according to Gorin (1950). To the south from those fractures the old mud cover had fallen by 3 meters. On the same volcano, according to Yakubov and Putkaradze (1950), from the east side of the crater a subsidence fracture of 500 meter length formed. After that the east part of the crater field subsided by 15-20 meters. In the north-east part of Kelany mud volcano, after the eruption of 1969, there was a fractured area of width 120-130 meters. Vertical displacement reached 2.5 meters and one of the fractures extended for 2-3 km (Yakubov et al., 1970).

On the Lokbatan volcano, after an eruption in 1951, fractures of 1 m width formed in an area of 400 meter width (Zhemerev, 1954). Two very long (about 1 km length) faults of 1.5-2 m displacement can be observed on this volcano today. Those faults were formed after the eruption of 1977. Fractures have been recorded on Bozdag Kobiiski (Yakubov et al., 1953), Melikchobanly (Zeinalov et al., 1969), Cheildag (Yakubov et al., 1972), Bulla Island (Sultanov et al., 1967) and a number of other volcanoes.

However, to measure statistically fracture-formation occurrence during an eruption around a mud volcano is not easy. There are no data available on subaqueous fractures, thereby restricting information on fractures to onshore volcanoes. However, even then, there are difficulties. In the majority of cases information on fractures is absent, and it is impossible to know whether the mud-volcano eruptions had or had not given rise to fractures, or whether fractures had occurred but no record was made. In addition, both radial and concentric (circular) desiccation fractures form as the mass of mud erupted from a volcano dries. These fractures are not of interest because they present no hazard and would not form in subaqueous conditions anyway. Literature sources often mention fractures associated with mud volcanoes, without specifying whether the fractures are eruption-derived or desiccation-derived.

Nonetheless, collected data on fractures from 57 mud volcanoes are presented in Table 3.1.

Table 3.1. Mud Volcano Fractures

1	2	3	4	5	6	7	8	9
1.	Lokbatan	19	10	9	2	1	0	0
2.	Shikhzagirli	16	10	6	2	2	1	0
3.	Keyreki	12	11	1	0	1	0	0
4.	Kushchi	12	10	2	0	1	0	1
5.	Bulla Island	9	6	3	1	0	0	0
6.	Bahar	8	8	0	0	0	0	0
7.	Touragai	7	6	1	1	1	0	0
8.	Nardaran-Akhtarma	7	6	1	0	0	0	0
9.	Bozdag-Kobi	6	4	2	1	1	1	0
10.	Otmanbozdag	6	2	4	3	1	1	1
11.	Dashgil	5	5	0	0	0	0	0
12.	Airantekyan	5	3	2	0	1	1	0
13.	Goturdag	4	3	1	0	0	0	0
14.	Dashmardan	4	0	4	2	2	2	1
15.	Akhtarma-Pashaly	4	4	0	0	0	0	0
16.	Shekikhan	4	4	0	0	0	0	0
17.	Akhtarma-Puta	3	3	0	0	0	0	0
18.	Bozdag-Guzdek	3	3	0	0	0	0	0
19.	Agnour	3	1	2	2	1	0	0
20.	Ayazakhtarma	3	3	0	0	0	0	0
21.	Kichik Maraza	3	3	0	0	0	0	0
22.	Saryncha	2	1	1	1	0	0	0
23.	Delyaniz	2	2	0	0	0	0	0
24.	Demirchi	2	2	0	0	0	0	0
25.	Karakure	2	1	1	0	0	0	0
26.	Kelany	2	1	1	0	0	0	0
27.	Chapylmysh	2	2	0	0	0	0	0
28.	Cheildag	2	1	1	1	1	0	1
29.	Jairli	2	1	1	0	0	0	0
30.	Melikchobanly	2	0	2	1	1	0	1
31.	Bozdag-Gekmali	1	0	1	0	1	0	0
32.	Adsyz	1	0	1	1	0	0	0
33.	Ajiveli	1	1	0	0	0	0	0
34.	Beyuk Kanizadag	1	0	1	0	0	0	0
35.	Durandag	1	1	0	0	0	0	0
36.	Zeiva	1	1	0	0	0	0	0
37.	Agzybir	1	0	1	1	0	0	0
38.	Bandovan	1	1	0	0	0	0	0
39.	Duzdag	1	1	0	0	0	0	0
40.	Kichik Kharami	1	1	0	0	0	0	0
41.	Neftchala	1	1	0	0	0	0	0
42.	Hamamdag	1	0	1	1	1	1	0
43.	SE Davalidag	1	0	1	1	1	1	0
44.	Shetanud	1	1	0	0	0	0	0

Table 3.1 (Continued)

1	2	3	4	5	6	7	8	9
45.	"New formed volcano"	1	1	0	0	0	0	0
46.	Akhtimer	1	1	0	0	0	0	0
47.	West Akhtimer	1	1	0	0	0	0	0
48.	North Shorsulu	1	1	0	0	0	0	0
49.	South Solakhai	1	1	0	0	0	0	0
50.	Zaakhtarma	1	0	1	0	1	0	0
51.	Girde	1	1	0	0	0	0	0
52.	Beyukdag	1	1	0	0	0	0	0
53.	Big Soap Springs	1	1	0	0	0	0	0
54.	Karyja	1	0	1	0	0	0	0
55.	Central Oukh	1	1	0	0	0	0	0
56.	Gotur	1	1	0	0	0	0	0
57.	Central Solakhai	1	1	0	0	0	0	0

Notes to Column Identifications:

1. Number.
2. Name of the volcano.
3. Number of recorded eruptions.
4. Number of eruptions where fractures have not been mentioned.
5. Number of eruptions where the presence of the fractures is mentioned.
6. Number of eruptions where the presence of many fractures is mentioned.
7. Number of eruptions where fractures with length from 100 to 500 m are mentioned.
8. Number of eruptions where fractures with length from 500 m to 1 km are mentioned.
9. Number of eruptions associated with fractures longer than 1 km.

A fracture-factor was used in the study, representing quantitatively the proportion of the number of eruptions accompanied by fracture-formation out of the total number of eruptions on a given volcano. As in the study of the ignition factor, detailed in Chapter 4, for the majority of volcanoes only zero, one or two eruptions have been recorded. In those cases it is difficult to characterize the process of fracture formation because if there was a total of just one eruption (and such mud volcanoes are in the majority) then the fracture-factor has the value 1 or 0, depending on whether a fracture was recorded at that volcano or not; the majority of fracture-factor values fall at the zero and unity ends of the interval of values. Therefore, only volcanoes were used in which five or more eruptions have been recorded. Because the fracture-factor varies in the range (0,1), a beta-distribution was used to examine the probability of fracture-formation. The characteristics of the beta-distribution are described in Bagirov et al. (1996). Using the method of maximum likelihood, the parameters of the probability distribution may be taken as $\alpha = 0.94$ and $\beta = 2.7$. The corresponding probability distribution density is presented in Fig. 3.6.

Unfortunately, the low number of data points did not permit a quantitative investigation of other factors associated with fracture formation. The process of fracture formation may depend on the volume of breccia erupted, because a large volume of breccia creates a high pressure on the rocks, which may result in destruction of the integrity of the formations. Three classes of breccia volumes and areas of cover for eruptions can be identified:

1. Those in which there were large fractures;
2. Those in which there were small fractures;
3. Those in which no fractures were recorded.

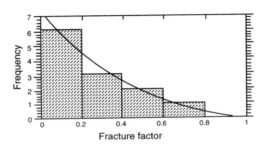

Figure 3.6. Beta-distribution of the fracture-factor with $\alpha=0.95$, $\beta=2.7$.

Statistical characteristics and distribution histograms of these classes are given in Figs. 3.7 and 3.8. Superficially these classes appear not to differ, although data sparsity is a major problem. The effect of breccia on fracture formation should be studied further with a larger data base of better quality.

D. Conclusions

There is some sort of relationship between seismic activity and mud-volcano activity. However, further study needs to be undertaken with a larger, higher quality set of data in order to sharpen the correlation.

Eruptions are often accompanied by the formation of fractures, and the lengths of fractures may reach several kilometers. The fracture-factor, introduced to assess the probability that fractures will form, is described by a beta-distribution with parameters $\alpha=0.94$ and $\beta=2.7$. Additional statistical analysis of the dimensions of fractures, and the probability that fractures of particular dimensions will form, together with a study of the volumes of breccia and the strength of the eruption on the fractures, did not yield results due to both the poor quality and sparseness of the data.

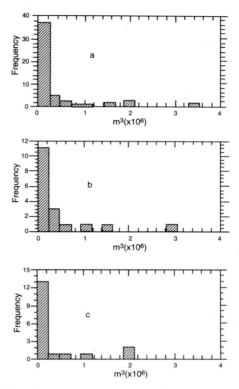

Figure 3.7. Distribution of volumes of breccia for different classes of eruption.

The work presented here, and elsewhere (Bagirov et al., 1996), has focused on aspects of mud volcanism and seismic activity in relation to hazards posed to offshore rig platforms. However, in trying to solve some problems, such as estimating the probability of gryphon-formation, or of fracture-formation during an eruption, there were insufficient satisfactory data. In association with an improved set of data comes the need to improve the quantitative methods.

For instance, in evaluating the average for a set of data in which some data points are missing, it was shown that multi-parameter distributions satisfied exponential conditions. However, other multi-parameter distributions (such as the gamma-distribution or the Weibull distribution) may better describe the system . Only a better data-set will help here. Equally, it would then be appropriate to include a quantitative coefficient describing the remoteness of a mud volcano from centers of population, which influences the record of eruptions. Again, a better data-set is needed to evaluate this concern.

Problems concerning the mutual interaction of mud-volcano activity and seismicity are also of interest; here the paucity of currently available data is particularly hard felt. Good, high quality, seismic and mud volcano fracture data are urgently needed.

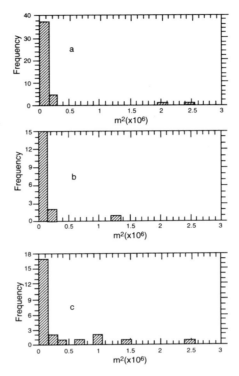

Figure 3.8. Distribution of area of cover for different classes of eruption.

Appendix A

Modeling strain

The distortions and rotations are indicated by changes in the shape and orientations of the sediment units. The geometries are known in the undeformed state and in each of the modeled time-steps. Because the sediments move around a changing mud diapir shape, the overall changes in displacements are not expected to be constant, i.e. the deformation is characterized as inhomogeneous. The individual sediment volumes may provide a different deformation history, i.e. the volumes may be interpreted as having undergone homogeneous deformation through time because the surfaces of the sediment volumes might be approximated as planar layers. Such an approximation is valid as long as the sediments are deformed on a scale small compared to the size of the sediment volumes being considered.

In three dimensions deformation of a solid requires specification of nine components of the strain tensor; the same number is also required for specifying the stress tensor. This number can be reduced to six by symmetry considerations. In two dimensions the specification of four components of strain can likewise be reduced to three. Deformation of the sediments is assumed not to result in any change in the volume of the solid (whereas the void space is allowed to alter during burial) but only to a change of shape that may include extensional, compressional and shearing components. Numerically the lateral and vertical velocities are known at each time-interval by mapping the paths and bed locations.

The strain induced in the sediments during mud diapir emplacement can be calculated at different locations by dividing the sediments into small cells. The boundaries of the cells are determined by the number of subdivisions specified for the individual formations and by the chosen spacing of the grid lines, respectively. By following the deformation of cells in time the deformation pattern can be deduced, and then, by comparing the displacement of cells defined and bounded by the same bedding surfaces, the development of strain can be mapped.

Once the strains for each cell are calculated the strain components can be rotated into the principal strain axis coordinate system where shear strain is zero. The representation of the fractional change in dimensions of a cell will be given by the principal strains.

The sign convention used for the stress and strain calculations is as follows: (a) compressive stress applied to a plane is considered positive, thus tensile stress is negative; (b) the normal components of strain are assumed to be positive if the deformation shortens the length of a side - a convention consistent with (1).

Modeling stress

Once the distribution of strain is calculated, the orientations of the principal axes of strain through time provide information on the stress causing the strain. However, the principal axes of strain will, in general, not coincide with the principal axes of stress because the strain is determined from the displacements, whereas the stress depends on the rates at which strains occur instead of strains themselves (Hobbs et al., 1976). A set of identical shape changes can lead to different stresses depending on material properties or different strain rates in the same material; the strain remains the same.

Sediments respond to an imposed stress system in a variety of ways depending on the physical conditions of the sediments. Even for sediments of the same type, the chemical and physical properties may change during burial, for example due to pressure dissolution and compaction. In order to characterize the response of rocks to stress, different classes of response such as elastic, viscous and plastic models are used because materials often display these types of responses over a variety of physical conditions. Depending on the scale of a sample of sediment, the sediment may be treated as being mechanically homogeneous or inhomogeneous. Sediments in the quantitative model are considered statistically homogeneous, i.e. all sediment volumes have identical mechanical properties within the limits of the size of the sediment volumes considered. The homogeneous material may be mechanically isotropic or anisotropic, where an anisotropic material is one in which the mechanical properties are dependent on the direction in which they are measured. A sediment like shale is highly anisotropic in strength due to the plate-like organization of the sediment particles.

When a sound wave travels through the bedrock the rock particles are locally displaced, but once the sound wave has passed, the particles return to their prior positions. The strain takes place within a very short period of time and is recoverable, i.e. there is no permanent distortion of the rock and the response to the disturbance is _elastic_. A linear relationship between stress and strain relates to materials that are considered isotropic and elastic and that have no preferred orientations of their mechanical properties.

Modeling fracturing of sediments

After calculation of the orientation and magnitude of the principal stress axes one can calculate the normal stress and shear stress; the relation between them can be illustrated by using the Mohr diagrams (Mohr, 1900; Lerche and Petersen, 1995). However, not all states of stress can be represented by a circle in the Mohr diagram because, for instance, the shearing stress along a plane may exceed some critical value and the material may undergo shear failure. The critical shearing stress can be functionally related to the stress normal to that surface. In the case where the critical shearing stress depends linearly on the normal stress, the Coulomb criterion can be used for shear failure, calculated through inherent shear strength of the material and the coefficient of internal friction.

Laboratory experiments have yielded a broad range of estimates of the uniaxial compressive strength of sedimentary rocks. Jaeger and Cook (1976) list a range of values from $1.4 \cdot 10^7$Pa to $3.5 \cdot 10^8$Pa. The coefficient of internal friction likewise ranges in value between 0.51 and 0.75, with an average value being around 0.65, which corresponds to an orientation of the faulting planes at an angle of approximately 28° to the maximum principal stress axis. Technical details describing how to implement these aspects of the calculations of evolving sediment properties are displayed in Lerche and Petersen (1995).

4 Mud Volcano Eruptions and Flame Hazards

A. Statistical Hazard Assessment from the Historical Record

From almost prehistoric times, records are available of eruptions of mud volcanoes associated with gas ignition, high flame, ejection of masses of mud, thrown masses of rock (sometimes of significant size), explosions and ground-motion trembles. In old manuscripts one can find descriptions of such natural phenomena. From the last century more or less periodic observations have been recorded in journals and guidebooks. By the middle of the 19th century geologists came to the conclusion that those explosions were associated with mud volcanoes (Abikh, 1939; Sjögren, 1887 and 1885-87). More and more facts on volcanic eruptions in this area were published. Later, mud volcanoes became, and still are, the objects of special attention of the Geological Institute of Azerbaijan. Field trips were arranged every year in order to observe volcanic activity. Hundreds of papers were published devoted to the analysis of activity and eruptions of specific volcanoes.

The purpose of this part of this chapter is to use almost two hundred years of recorded information to build a statistical pattern for assessing risk due to flaming eruptions in the volcanic mud region of Azerbaijan. The related seismic and fracture problems have been discussed in previous chapters.

Missing data are one of the biggest concerns when doing data analysis. In contrast to earthquakes, where we have almost continuous records, mud volcano eruptions are not only unpredictable in the same way as earthquakes, but are very local and so could easily be missed - unlike the regional effects of earthquakes. Most of the eruptions have not been recorded. The number of missing events has decreased with time towards the present. In our constructions we used information starting from 1810. At those time just a few very strong eruptions were noted. In 1970, the Geological Institute organized annual expeditions to all known volcanoes. So the probability of "missing" eruptions after those years is very small. Another factor related to the probability of "missing" eruptions is the "power". The more spectacular an eruption, so the less chance that it was missed. All of those factors have been accounted for during analysis of the statistical data. In order to illustrate the methods used, a specific area, Chirag, was chosen for consideration. But similar constructions can be done for any areas and volcanoes.

1. Descriptive Information

Mud volcanoes are geological bodies on the earth's surface composed of sedimentary rocks carried to the surface from deeper levels.

The mechanisms of mud volcanism are not yet completely clear. However, the known connections with mud diapirs and with deep layers (10-20 km) of the sediment pile suggest very complex mechanisms of development. Hydrocarbon gases nearly always occur with mud-volcano activity, suggesting they are a driving force of mud-volcano activity. In a later chapter a model of mud diapiric and volcanic activity will be considered incorporating gas-drive. But here we discuss only statistical inferences deducible from the historical record.

Mud volcanoes tend to occur on young platforms with thick sedimentary cover, and are generally associated with oil and gas. Geographically, mud volcanoes occur in Italy, Romania, Russia (the Kerch and Taman peninsulas, and Sakhalin), Turkmenistan, Georgia, Iran, Pakistan, India, Burma, China, Japan, Indonesia, Malaysia, Mexico, New Zealand, Colombia and Trinidad. However, Azerbaijan and the adjacent Caspian Sea is the classic region for mud volcanoes, where more than a third of the mud volcanoes of the world are concentrated (more than 250).

Morphologically mud volcanoes occur as hills and raised areas, sometimes reaching upward of 400 m in height, and with volumes of up to several x 10^7 m^3. Externally, mud volcanoes are similar to magmatic volcanoes, often displaying a dome-like structure. The products of mud-volcano activity are carried to the surface along exit channels, leading to craters at the surface. The crater field is most commonly circular to oval in outline and is surrounded by one or more concentric crater ramparts. The crater forms an area of subsidence, and varies in form from gently convex to a deep caldera. The area of the crater plateau can reach 10 km^2, and the crater rampart may rise 5-25 m above the center of the crater. The morphology of mud-volcano breccia flows depends mainly on the topography and on the breccia composition.

Volcanoes lying offshore differ morphologically from those on land, forming islands (7 such islands occur in the Caspian Sea) or submarine banks. At the time of an eruption some volcanoes form new islands, which are eroded within several days.

Mud volcano activity can be divided into two stages: gryphon-salse and eruptive. The first stage involves a gentle emission of relatively small quantities of water, mud and gas from gryphons, domes and salses lying on the mud volcano. Some gryphons emit oil, or an oily film, onto the surface of the water.

Gryphons form small conical geological bodies resulting from volcanic mud emission. Mud containing gas bubbles forces its way through the central channel of the gryphon to its peak and flows down the sides of the gryphon. The dimensions of gryphons vary from a few centimeters to meters. Large gryphons, known as domes, are produced when a closely spaced group of active gryphons combine to form a single gryphon, and can reach dimensions of multiples of meters. The formation of gryphons and domes is affected by the ratio of solid to liquid eruption products in the mud. When water-bearing flows contain only a small volumetric fraction of clay minerals, salses form as lakes filled with mineralized water, reaching dimensions of several tens of meters. The gryphon-salse stage is interrupted by eruptions when, over several hours or days, massive amounts of breccia and gas are ejected. Breccia flows, several meters thick, spread over hundreds of meters (occasionally several kilometers) in length. The gas emitted often ignites, and the flame height can reach hundreds of meters. During an eruption, fissures up to a meter or more wide may form, with offsets of several meters and lengths of several kilometers.

These factors must be included when planning risk assessment for rigs in mud volcano areas. For Azerbaijan, parameters related to volcanic eruptions have been collected from published sources, as presented by Yakubov et al. (1974) and Rakhmanov (1987). The data are summarized in Table 4.1.

The calculation of the <u>area</u> of cover of an eruption, given the length and width of flow and the average thickness, can easily be made. In calculating the <u>volume</u>, V, of breccia in submarine eruptions the formula used is for the volume of a truncated cone yielding:

$$V = (1/3)\pi H(R_1^2 + R_1 R_2 + R_2^2)$$

where H is the water depth; R_1 is the radius of the island; and R_2 is the radius of the submarine base of the island, and where the submarine slope angle of the volcano, θ, is taken to have $\tan\theta \approx 1/6$.

Because the radius over which fragments of rock were scattered has been recorded in only a very few volcanoes, we do not use that information here. However, this information is not wasted because it can be used for calculating the height of the mud eruption as follows: The maximum range of a physical body discharged at a fixed velocity is reached at a discharge angle of $45°$ from the horizontal (ignoring friction). The maximum range is four times the maximum height. The Kelany volcano (in 1969) provides the single case of both the height and radius being recorded over which fragments were scattered and fits the range/height ratio of 4/1 nicely. Therefore, data gaps in the heights of eruption have been filled for those eruptions where the collection radius of scattered fragments was recorded.

2. Onshore versus Offshore Statistics

In assessments of the risk of safe operations of offshore installations it is important to investigate not only the probability of a known mud volcano erupting within the field-development area, but also of the formation of new mud volcanoes. To study the probability of new mud volcanoes forming, all of the available information on mud volcanoes was analyzed (Table 4.1) with the broad-based conclusions that:

1. The number of recorded mud-volcano eruptions on land exceeds, by nearly two orders of magnitude, the number of recorded eruptions offshore.

Table 4.1. (cont.)

Eruption Number	Number of eruptions per volcano	Date of eruption	Years since the eruption at 01.01.93	Time since last eruption (yrs)	Length of tongue of erupted breccia (m)	Width of tongue of erupted breccia (m)	Equivalent circular diameter of erupted breccia (m)	Area covered (10^6 m^2)	Thickness of cover (m)	Volume of breccia (10^6 m^3)	Height of island (m)	Length of island (m)	Width of island (m)	Diameter of island (m)	Water depth around island (m)	Height of eruption (m)	Radius of area covered by scattered fragments (m)	Flame height (m)	Time length of Burn	Duration of eruption	Fractures	Noise r-rumble e-explosion h-hissing t-tremor
34	15	1991Apr.	1.7	4.0			30-35	0.8	0.4	0.3								none				r
35	16	1992Sep.	0.3	1.4			25-30	0.5	0.5	0.25								none				
K E Y R E K I																						
36	1	1830May	162.6	-														+		more than a day		e
37	2	1865	128	34.6																		
38	3	1885	108	20																		
39	4	1902	91	17						+												
40	5	1915Feb.	77.8	13.2																5 hrs.		r
41	6	1952Aug.	40.3	37.5				300	0.5-2	400-600						30-40				4.5 hrs.		e,r
42	7	1953	40	0.3																		
43	8	1964	29	11																		
44	9	1966July	26.4	2.6	150-30+1															2 hrs.		
45	10	1968Apr.	24.7	1.7	285	55		18.7	0.5-1	15						20-25				3 hrs.		
46	11	1989Feb.	3.8	20.9	100+70					+						5-10						r
47	12	1991Sep.	1.3	2.5												5-7						r
K U S H C H I																						
48	1	1917Nov.	75.1	6.1						+								200	2 hrs.			
49	2	1924	69.0	17														100-200				
50	3	1941	52	17														100-200				
51	4	1952	41	11																		
52	5	1954	39	2																		
53	6	1958	35	4																		
54	7	1960Aug.	32.3	2.7																		
55	8	1961	32	0.3																		
56	9	1965Sep.	27.3	7.7				6.3	1-1.5	4-6						15-20		+	2 hrs.			r
57	10	1986Mar.	6.8	20.5	120	115		10	0.5	2.4						80-120		+	1 hr.		length 100 m	r
58	11	1992Oct.	0.2	6.6	80	60		4.8	0.5							100		+				
M A K A R O V A B A N K																						
59	1	1876	117	30					944		10											
60	2	1906	87	6																		
61	3	1912	81	5																		
62	4	1917	76	4																		
63	5	1921	72	4.4					927			island	small									
64	6	1925July	67.6	4.4					1500	5			600	20								
65	7	1933	60	7.6																		
66	8	1941	52	8																		
67	9	1958Oct.	34.2	17.8														200-250	30 min.	40 min.		e

Table 4.1. - Parameters of Mud Volcano Eruptions

LOKBATAN (Eruptions 1–19), **SHIKHZAGIRLI** (Eruptions 20–33)

Eruption Number	Number of eruptions per volcano	Date of eruption	Years since the eruption at 01.01.93	Time since last eruption (yrs)	Length of tongue of erupted breccia (m)	Width of tongue of erupted breccia (m)	Equivalent circular diameter of erupted breccia (m)	Area covered (10^6 m^2)	Thickness of cover (m)	Volume of breccia (10^6 m^3)	Height of island (m)	Length of island (m)	Width of island (m)	Diameter of island (m)	Water depth around island (m)	Height of eruption (m)	Radius of area covered by scattered fragments (m)	Flame height (m)	Time length of Burn	Duration of eruption	Fractures	Noise (r-rumble, e-explosion, h-hissing, t-tremor)
1	1	1828 Dec.	164.0	-																		
2	2	1864 May	128.6	35.4																		
3	3	1887 Jan.	105.9	22.7	300	200		60	2	120								600	3-4 hrs	10 days	many	e
4	4	1890	103	2.9																		
5	5	1904	89	14.0																		r
6	6	1915 Jan.	77.9	11.1														60	1 hour	1 hour	+	
7	7	1918 Mar.	74.8	3.1																		
8	8	1923 Jan.	69.9	4.9				5										40				
9	9	1926 July	66.4	3.5	153	80		12	1	12						25	100	40/60	10min/1	3.5 hrs	2-3m deep	
10	10	1933 Mar.	59.8	6.6	300																	
11	11	1935 Feb.	57.8	2.0	250	100		25												6-8 days	+	
12	12	1938 Jan.	54.9	2.9						little												
13	13	1941 Mar.	51.8	3.1	120	30		3.6													+	r,e
14	14	1954 July	38.4	13.4	120	30		41.6	1-5	104						25	100	450/10-15	15 min.	2 days	+	r,e,h
15	15	1959 Dec.	33.0	5.4				100	0.5-3	170						40-50		200-300	10 min.	2 days	none	r,e,t
16	16	1972 Oct.	20.2	12.8	100-200	50-60		6-10										200-300	3 min.			r,t
17	17	1977 Oct.	15.2	5.0	210	50-120		55	4	200								200-300	5 times	2 days	+	
18	18	1980 Mar.	12.8	2.4	210	50-120	in the cent 80	20-25										0	0		-	
19	19	1990 Mar.	2.8	10.0						little								0	0		+	
20	1	1846 July	146.4	-														+				
21	2	1868 May	124.6	21.8														+				
22	3	1872 Jan.	120.9	3.7																		
23	4	1902 Jan.	90.9	30						much								black smoke-100	3 hours	48 hrs.		e
24	5	1927 Aug.	65.3	25.6	115		110	10	1.5-2	15-20								200-300			width=1m; in places = 12 m	r,e,t
25	6	1929 Nov.	63.1	2.2			110	8	1-2	10-12						75	300	200-300				r,e,t
26	7	1939 Mar.	53.8	9.3			160-200	20-30								100		+				
27	8	1946 July	46.4	7.4																10 days	width-0.5m throw-0.7m	
28	9	1949 April	43.7	2.7			80	5	0.5-2	6-8								+				
29	10	1955 Jan.	37.9	5.8														100-200			+	
30	11	1969 Sum.	23.4	14.5														+				r
31	12	1977 Sum.	15.4	8.0	100	100		9	1.5-2	20								100				r
32	13	1980 Nov.	12.1	3.3			100	8	1-2	10-12								100	2 hours		2, big	r
33	14	1987 Apr.	5.7	6.4					1									100				r

Table. 4.1 (cont.)

Eruption Number	Number of eruptions per volcano	Date of eruption	Years since the eruption at 01.01.93	Time since last eruption (yrs)	Length of tongue of erupted breccia (m)	Width of tongue of erupted breccia (m)	Equivalent circular diameter of erupted breccia (m)	Area covered $(10^6 m^2)$	Thickness of cover (m)	Volume of breccia $(10^6 m^3)$	Height of island (m)	Length of island (m)	Width of island (m)	Diameter of island (m)	Water depth around island (m)	Height of eruption (m)	Radius of area covered by scattered fragments (m)	Flame height (m)	Time length of Burn	Duration of eruption	Fractures	Noise r-rumble e-explosion h-hissing t-tremor
68	10	1963Feb.	29.8	4.4												20*				30 min.		
69	11	1984May	8.6	21.2												15-20*						
B A H A R																						
*height of ejected water is indicated																						
70	1	1853	140																			
71	2	1859	134	6																		
72	3	1886July	106.4	27.6														+		20 days		
73	4	1909Mar.	83.8	22.6																9 days		
74	5	1911	82	1.8																		
75	6	1926Oct.	66.2	15.8														+				
76	7	1967Sep.	25.7	40.5			200	2	400									100-200			very long	e
77	8	1992Oct.	0.2	25.5			80	2	160									500		2 days		e
B U L L A I S L A N D																						
78	1	1857Mar.	135.8		359					16								+	45 min.			r,e
79	2	1859July	133.4	2.4						17.4	3.6	87	66		1.2			+				r
80	3	1940Aug.	52.3	81.1				1000+ 300		12.1000								1000/ less	15 min./ 50 min.	45 hrs.	width 1-2 m	
81	4	1947Aug.	45.3	7.0												2-3		none				
82	5	1959July	33.4	11.9	150	100		15	0.3	5											width 1 m	
83	6	1960July	32.4	1.0	500	450		225	1.5	270						200				4 hrs.	width 1-3m	r,e
84	7	1963Mar.	29.8	2.6																		
K U M A N I B A N K																						
85	1	1861Mar.	131.6					2.8		18	0.6	73	55		5.5	0.5m*						
86	2	1927May	65.6	66						350	1.5			350								
87	3	1927Nov.	65.1	0.5						240	3.4			288								r
88	4	1928Nov.	64.1	1.0		50		65		1500	3	324	230							7 min.		
89	5	1939Nov.	53.1	11.0						1160	6	700	500					100	15 min.	30 min.	+	
90	6	1950Dec.	42.0	11.1														200	20 min.			
91	7	1959Dec.	33.0	9.0						240	2.5	200	170			3-4m*						r,e
T O U R A G A I																						
92	1	1841	152																			
93	2	1901	92	60																		
94	3	1924Mar.	68.8	23.2	530	235		125										+				
95	4	1932Apr.	60.7	8.1														+				r
96	5	1947Nov.	45.1	15.6												500				30 min.	up to 500m	
97	6	1950	43	2.1														500				
98	7	1985	8	35			140	15	0.5	8												

Table 4.1. (cont.)

Eruption Number	Number of eruptions per volcano	Date of eruption	Years since the eruption at 01.01.93	Time since last eruption (yrs)	Length of tongue of erupted breccia (m)	Width of tongue of erupted breccia (m)	Equivalent circular diameter of erupted breccia (m)	Area covered $(10^6 m^2)$	Thickness of cover (m)	Volume of breccia $(10^6 m^3)$	Height of island (m)	Length of island (m)	Width of island (m)	Diameter of island (m)	Water depth around island (m)	Height of eruption (m)	Radius of area covered by scattered fragments (m)	Flame height (m)	Time length of Burn	Duration of eruption	Fractures	Noise (r-rumble, e-explosion, h-hissing, t-tremor)
		NARDARAN AKHTARMA																				
99	1	1948Nov.	44.6																			
100	2	1971	22	22.6																		
101	3	1972	21.5	0.5																		
102	4	1972	21	0.5																		
103	5	1982	11	10	200	150		30	1.5	45								5-6				
104	6	1984	9	2	200	180		36	1	36												
105	7	1986	7	2	300	20		60	1.2	72											+	
		BUZOVNY DOME																				
106	1	1892May	100.6							25	8.5	51	32		12							
107	2	1915	78.0	22.6																		
108	3	1923Oct.	69.2	8.8						17	3	S = $1610 m^2$			5.6	20*		+				
109	4	1950July	42.4	26.8						30	3-4	60	90									
110	5	1953Feb.	39.8	2.6						50	3.5	60	50			15-30*						
111	6	1953July	39.4	0.4						30-35	5	60	90			10-15*						
		OTMANBOZDAG																				
112	1	1854Jan.	138.9				60-65	3	2	6						5						
113	2	1904Nov.	88.1	50.8				2500										+		3 hrs.	width = 20 cm	e
114	3	1922Jan.	70.1	18	300	230		1500	1-5	210											length = 1200m width = 0.7m	e
115	4	1951	42	38.1				70	1-2	170								100-200			many	
116	5	1965Oct.	27.2	14.8			230	50										100	2 hrs.		throw = 0.5 m	r
117	6	1985	7	20.2				3		+								+				
		BOZDAG KOBI																				
118	1	1827Nov.	165.1		600	400		240	1	240								very high to 1 m	3 hrs./24 hrs.			
119	2	1894	99	66.1																		
120	3	1902	91	8																		
121	4	1953Aug.	39.3	51.7	150	10		15	0.22	0.04								10 m	4 days		+	r
122	5	1957Aug.	35.3	4.0				60	5	9												
123	6	1974May	18.6	16.7	300	200				310						400-500		300	1 hr.		+	r
124	7	1987	6	12.6			60	3		+												
		DASHGIL																				
125	1	1882	111																			
126	2	1902	91	20														+			+	
127	3	1908	85	6																		
128	4	1926	67	18																		

Table 4.1. (cont.)

Eruption Number	Number of eruptions per volcano	Date of eruption	Years since the eruption at 01.01.93	Time since last eruption (yrs)	Length of tongue of erupted breccia (m)	Width of tongue of erupted breccia (m)	Equivalent circular diameter of erupted breccia (m)	Area covered (10^6 m²)	Thickness of cover (m)	Volume of breccia (10^6 m³)	Height of island (m)	Length of island (m)	Width of island (m)	Diameter of island (m)	Water depth around island (m)	Height of eruption (m)	Radius of area covered by scattered fragments (m)	Flame height (m)	Time length of Burn	Duration of eruption	Fractures	Noise r-rumble e-explosion h-hissing t-tremor
129	5	1958	35	32														+				r,e
AIRANTEKYAN																						
130	1	1964	28		400-500	200-250												100-150	20-30 min.		+	r,e
131	2	1969	23	5	1600-1800	80-50		175 / 400-450	4 / 1	700 / 100						90		10-20 / 0.5-1 / 40	15 days		length=500 m width=2m depth=3-4m	e
132	3	1977	15	8				63	4	250								1000	5 hrs.			r,e
133	4	1988	4	11						+								50				
134	5	1990	2	2						+								100-120				e
AKHTARMA PASHALY																						
135	1	1948	45	21																		
136	2	1969	24	13						+												e
137	3	1982	11.1	13			220	38	1	38												
138	4	1982	11	0.1			130	13		+												e
KOTURDAG (dry extrusion (?))																						
139	1	1966	27		350	60-70		22	3-4	80-90											width 2-3m	
140	2	1970	23	4				8		+												
141	3	1977	16	7	300-350	15-35																
142	4	1987	6	10																		
GLINYANYI ISLAND																						
143	1	1810	183	50														+				
144	2	1860	133	53														110	0.5 hr.			r,e
145	3	1913	80	13														+				
146	4	1926	67						max:6-8									100				
147	5	1937	56	11				2000										1000	0.25			
LOS ISLAND																						
148	1	1876	117	47														+				
149	2	1923	70																			
150	3	1977	16	54				3/4 of island's area		60-70								1000				
POGORELAYA PLITA BANK																						
151	4	1993	0	16																		
152	1	1813	180																			
153	2	1825	168	12						3330	20	1000	750									
154	3	1843	150	18						1900	20	1000	750									

Table 4.1. (cont.)

Eruption Number	Number of eruptions per volcano	Date of eruption	Years since the eruption at 01.01.93	Time since last eruption (yrs)	Length of tongue of erupted breccia (m)	Width of tongue of erupted breccia (m)	Equivalent circular diameter of erupted breccia (m)	Area covered (10^6 m²)	Thickness of cover (m)	Volume of breccia (10^6 m³)	Height of island (m)	Length of island (m)	Width of island (m)	Diameter of island (m)	Water depth around island (m)	Height of eruption (m)	Radius of area covered by scattered fragments (m)	Flame height (m)	Time length of Burn	Duration of eruption	Fractures	Noise r-rumble e-explosion h-hissing t-tremor
											S V I N O I I S L A N D											
155	4	1868	125	25																		
156	1	1923	70																			
157	2	1927	66	4					18	1	$S=2800\ m^2$			4.5 m				300	13 min.			
158	3	1932	61	5					617		$S=5800\ m^2$							100-200	9 hrs.			e
159	4	1933	60	1														5	4 hrs.			
											L I V A N O V A B A N K											
160	1	1895	98																			
161	2	1930	63	35	200					1	40	20				5-8 m*						
162	3	1937	56	7						3.5	40	25										
163	4	1982	11	45						2.5			47									
											K O R N I L O V & P A V L O V B A N K											
164	1	1907	86	8					1	-0.3	27	13										
165	2	1915	78	8					7	1.5	64	20			3							
166	3	1970	23	55					55	6	114	45			5							
											A K H T A R M A P U T A											
167	1	1923	70																			
168	2	1933	60	10					+	+												
169	3	1950	43	17																		
											B O Z D A G - G U Z D E K											
170	1	1839	154	63														+	20 hrs.			e
171	2	1902	91																			
172	3	1988	5	86			100	8	1	8												
											A G N O U R											
173	1	1940	53							little						20-30		300	30 min.			
174	2	1948	45	8						little						20-30		200				
175	3	1976	17	28				50	3	150												
											A Y A Z A K H T A R M A											
176	1	1927	66																			
177	2	1973	20	66			200	30	0.3-0.5	12							0	0			+	
178	3	1985	8	12																	+	r,e
											M A L O M A R A Z A											
179	1	1969	24																			
180	2	1970	23	1												10	40					
181	3	1987	6	17			80	5														

Table 4.1. (cont.)

Eruption Number	Number of eruptions per volcano	Date of eruption	Years since the eruption at 01.01.93	Time since last eruption (yrs)	Length of tongue of erupted breccia (m)	Width of tongue of erupted breccia (m)	Equivalent circular diameter of erupted breccia (m)	Area covered (10^6 m^2)	Thickness of cover (m)	Volume of breccia (10^6 m^3)	Height of island (m)	Length of island (m)	Width of island (m)	Diameter of island (m)	Water depth around island (m)	Height of eruption (m)	Radius of area covered by scattered fragments (m)	Flame height (m)	Time length of Burn	Duration of eruption	Fractures	Noise r-rumble e-explosion h-hissing t-tremor
										S A R Y N C H A												
182	1	1936	57		300-350 / 200-250	100-180 / 100-150		60-65	2	120-130											+	
183	2	1976	17	40	400	40-60		20														
										D E L A N I Z												
184	1	1912	81													great height				20 days		e
185	2	1951	42	39					+													
										D E M I R C H I												
186	1	1958	35																			
187	2	1971	22	13			100-120	10								0		0				
										K A R A K Y U R E												
188	1	1928	65																			
189	2	1951	42	23	270	22		60	0.5-2	75								20-25	35 min.		length = 600 m	r
										M E L I K C H O B A N L Y												
190	1	1967	26					30	0.3-2.5	35-40						150				1 hr.	length = 100 m	r
191	2	1977	16	10			90	6								25				10 min.	length = 1000 m	
										H A M A M D A G												
192	1	1947	46				400	160	1	160											length = 1200	r,e
193	2	1984	9	37			140	16	1	16								+			major	
										K E L A N Y												
194	1	1962	31		180	160		29	0.5-1	160								400-800	3 hrs.			r
195	2	1969	24	7	500	350		70	3-8	500						150-200	1000	350-400	5-15 mins. 6 times		length = 6 km	r,e
										C H A P Y L M Y S H												
196	1	1929	64					2500										+				r,t
197	2	1933	60	4														100				
										C H E I L D A G (dry extrusion)												
198	1	1970	123		60	50		3	6	20								+			length 1km	e
199	2	1970	23	100																	many small	

Table 4.1. (cont.)

Eruption Number	Number of eruptions per volcano	Date of eruption	Years since the eruption at 01.01.93	Time since last eruption (yrs)	Length of tongue of erupted breccia (m)	Width of tongue of erupted breccia (m)	Equivalent circular diameter of erupted breccia (m)	Area covered ($10^6 m^2$)	Thickness of cover (m)	Volume of breccia ($10^6 m^3$)	Height of island (m)	Length of island (m)	Width of island (m)	Diameter of island (m)	Water depth around island (m)	Height of eruption (m)	Radius of area covered by scattered fragments (m)	Flame height (m)	Time length of Burn	Duration of eruption	Fractures	Noise r-rumble e-explosion h-hissing t-tremor
D A S H M A R D A N																						
	1	before 1854			1500	500-600		750	1-1.5	1000											+	
	2	before 1854			800-900	100-120		100	2	200											+	
200	3	1954	39					500	3-4	2000						300		600		1.25 hr.	throw= 2-5 m	r,e
201	4	1976	17	22														300				e
S H E K I K H A N																						
202	1	1988	5				150	20	0.5	10												
203	2	1989	4	1	150	100	100	8	0.5	4												
								15	1	15												
F E R S M A N										+												
P E R S I A N I N B A N K																						
204	1	1987 Apr.	56			80-100										1-1.5m*						
205	2	1987 Dec.	5	0.6												20-30m*						
J E I R L I																						
206	1	1868	125																			
207	2	1913	80	45																		
208	1	1951	42							+											length = 40 m width = 1-2 m	
N O R T H J E I R L I																						
209	1	1983	10		200	100		20														
B O Z D A G G E K M A L Y																						
210	1	1926	67		80	70	80	5	0.7	3										30 hrs.	2 cracks L = 40 m	
A D S Y Z																						
211	1	1949	44					5.6	0.6-2	7											length = 40-45 m width = 1.0-1.5 m	
D U R A N D A G																						
212	1	1960	32		700	300		210	3	630								100				
Z E I V A																						
213	1	1926	67																			
A G Z Y B I R																						
214	1	1964	29															0	0	0		r

Table 4.1. (cont.)

Eruption Number	Number of eruptions per volcano	Date of eruption	Years since the eruption at 01.01.93	Time since last eruption (yrs)	Length of tongue of erupted breccia (m)	Width of tongue of erupted breccia (m)	Equivalent circular diameter of erupted breccia (m)	Area covered (10^6 m^2)	Thickness of cover (m)	Volume of breccia (10^6 m^3)	Name	Height of island (m)	Length of island (m)	Width of island (m)	Diameter of island (m)	Water depth around island (m)	Height of eruption (m)	Radius of area covered by scattered fragments (m)	Flame height (m)	Time length of Burn	Duration of eruption	Fractures	Noise (r-rumble, e-explosion, h-hissing, t-tremor)
215	1	1932	60								BANDOVAN												
216	1	1941	52							+	DUZAG												
217	1	1912	80							+	MALYI KHARAMI												
218	1	1947	46							+	NEFTECHALA				50		+						
219	1	1954	38								AJIVELI												
220	1	1950	43		110-120				1.5	100	BOLSHOI KANIZADAG						70-80		150-200	1.5 hrs.	6 hrs.	+	r,e
221	1	1989	4				100	8	1	8	SOUTH SOLAKHAI												
222	1	1990	3				100	8	3	24	ZAAKHTARMA								500			+	r
223	1	1962	31		230	170			5	160	GIRDE								400	3 hrs.			r,e
224	1	1987	6								BOLSHIYE MYL'NYE RODNIKI												
225	1	1987	6		2 X 45						KARYJA												
226	1	1988	5							+	CENTRAL OUKH												
227	1	1983	10				70	4	0.5	2	GOTUR												
228	1	1989	4				200	31			CENTRAL SOLAKHAI												
229	1	1975	18				200	31			SOUTH-EAST DAVALIDAG						20-3		0	0	0.1 day	+	
230	1	1975	18					33	2	66	SHEITANUD								+			+	
231	1	1980	13								A NEW MUD VOLCANO BANK						15-20m*						
232	1	1985	8		300						AKHTIMER (AGDAM)												

Table 4.1. (cont.)

Eruption Number	Location	Number of eruptions per volcano	Date of eruption	Years since the eruption at 01.01.93	Time since last eruption (yrs)	Length of tongue of erupted breccia (m)	Width of tongue of erupted breccia (m)	Equivalent circular diameter of erupted breccia (m)	Area covered (10^6 m²)	Thickness of cover (m)	Volume of breccia (10^6 m³)	Height of island (m)	Length of island (m)	Width of island (m)	Diameter of island (m)	Water depth around island (m)	Height of eruption (m)	Radius of area covered by scattered fragments (m)	Flame height (m)	Time length of Burn	Duration of eruption	Fractures	Noise (r-rumble, e-explosion, h-hissing, t-tremor)
233	NEW FORMED MUD VOLCANO BANK	1	1981	12				5	0.02														
234	WEST AKHTIMER	1	1988	4				80	5	0.5	2.5												
235	NORTH SHORSULU	1	1986	7		60																	
236	KARAGEDOV BANK	1	1957	36																			
237	DUVANNY ISLAND	1	1961	32					200	10	2000						45		200-300	0.7 hrs.	45 min.		r,e
238	KAMNI IGNATIYA	1	1920																400	15 min.			
239	BEYUK DAG																						
240	12 MILES FROM SANGI MUGAN ISLAND																						

2. The volumes of the recorded offshore mud volcanoes eruptions are independent of water depth and are comparable with the volumes of the most powerful onshore eruptions.

3. As the water depth increases the number of recorded offshore mud-volcano eruptions decreases.

4. For those offshore volcanoes where only one eruption has been recorded, it is not possible to say whether this was a new mud volcano, since there is no guarantee that there was no earlier (or even subsequent) eruption, due to the paucity of studies of the sea-bed sediments.

The impoverished state of recorded offshore mud volcanoes forces an exclusion of the offshore data. Probability analysis for formation of a new mud volcano then relies on the data of higher quality (and quantity) from onshore mud volcanoes.

The study of mud volcanoes on land enables morphotectonic investigations to be carried out, trenches to be dug, and similar field work, thus guaranteeing identification of 100% of the mud volcanoes which are new over historical time. The absence of inherited morphology, traces of earlier breccia eruptions, transported mud-volcano conglomerates, and a number of other factors, enable identification of any land mud volcano which has been formed within the last million years. Of course, the possibility is not excluded that, over geological time extending for tens of millions of years, there has never been a previous mud-volcano eruption wherever a "new" mud volcano has appeared. Traces of ancient mud-volcano eruptions have been found in Azerbaijan in sections of deep wells dating from the Neogene, and even in places where mud volcanoes are now absent. However, all the mud volcanoes which were identified as "new" from geomorphological data, and on the basis of information on Quaternary geology, have been unexplored by deep drilling, and it was not possible to estimate the likelihood of mud-volcano activity over geological time. Therefore, the term "new" mud volcano means a mud volcano for which it is known, for certain, that there has been no previous eruption in the same place over the last million years. Study of the available data has enabled the recognition of several such new mud volcanoes onshore in Azerbaijan as listed in Table 4.2.

3. Probability of a New Volcano Developing

The off-shore Chirag area is one of considerable interest to oil companies. Therefore in this sub-section an estimate is given of the probability of a new mud volcano appearing in that area. Consider that the formation of new mud volcanoes occurs with equal probability on land and offshore. Knowing the number of new mud volcanoes, the area over which they formed, and the area under investigation (Chirag), together with the time interval, it is possible to calculate the probability that a new mud volcano will form in the Chirag area.

The area of mud volcano occurrences onshore in Azerbaijan comprises: Apsheron Peninsula of area 1.8×10^3 km^2; Shemakha-Gobustan province of area 6.7×10^3 km^2; Lower Kura province of area 9.1×10^3 km^2. The total onshore area affected by mud volcanism in Azerbaijan is, then, 17.6×10^3 km^2.

Table 4.2. Newly-formed mud volcanoes

No.	Volcano name	Year of Formation	Notes
1.	Beyuk-dag	Winter 1920-1921	Below Binagady settlement. Cone 1 m high, and 2.5 m in diameter.
2.	Bezymyannyi	23 October 1949	62 km NE of Karakyure. Covered an area of 80 x 70 m and 0.6-2.0 m high, including 7 salses and numerous fractures.
3.	Girde	Middle of September 1962	7 km SE of Kushchi mud volcano. Mass of breccia 350 thousand tons.
4.	New mud volcano	1981	Resulted from Ismailli earthquake, formed a volcano 5 m across.

Let the probability that a new mud volcano will form in the Chirag area on a certain day be p. Then the probability that k volcanoes will form during t years is described by the Poisson process:

$$P\{N(t) = k\} = \exp(-\lambda t)(\lambda t)^k/k! \tag{4.1}$$

with parameter $\lambda = 365p$.

Let $A_c = 270$ km^2 represent the area of the Chirag field (Yakubov et al., 1974); let $A_T = 17,560$ km^2 represent the total area of mud volcano development onshore in Azerbaijan; Then with k=4, the number of newly formed volcanoes over 100 years, we have

$$\lambda = 4A_c/(100A_T) = 6.2 \times 10^{-4} \text{ yrs}^{-1}.$$

The probability that not a single volcano will form over a period of t years is then just $P_o = \exp(-\lambda t)$; while the average waiting time for the appearance of a new mud volcano within the Chirag area is just $\tau_1 = 1/\lambda \cong 1600$ yrs.

4. Distance distribution of individual groups of gryphons from a common eruptive center

Analysis of data from mud volcanoes in Azerbaijan shows that one, or several, centers of gryphon formation occur. The statistics for the number of gryphon formation centers are given in Table 4.3, and the number appears to be independent of such factors as the total volume of emission from the mud volcano, cone height, etc.

Table 4.3. Distribution of number of centers of gryphon formation on mud volcanoes

Number of centers of gryphon formation	5	4	3	2	1	0
Number of volcanoes	2	5	8	2	25	4

Further, on large volcanoes, which form mainly as the result of eruptive material emission, both the number of gryphons and the number of gryphon formation centers is smaller than those with small and non-erupting volcanoes, where the mud emission comes mostly from gryphons and salses. In respect of gryphon formation, mud volcanoes can be divided into two types, central and scattered, with distance from a single eruptive center. Central volcanoes have conical, isometric structures, heights of 100 m or more and a significant crater, with a single gryphon formation center lying in the crater; examples include Touragai, Bolshoi Kyanizadag, and Tashmardan, reaching 400 m in height, and for which there is one gryphon formation center with a small number of intermittent gryphons. For central volcanoes the probability that the gryphon formation center will be distant from the eruptive center is limited by the crater size.

By way of contrast, the scattered type of volcano has several (up to five) separate gryphon formation centers which are several hundred meters to several kilometers apart. For example, Cheildag has four gryphon formation centers lying 1-2 km apart, Ajiveli has five gryphon formation centers with a maximum separation of 10 km. Gryphon formation centers are usually associated with a system of fractures and faults belonging to a single zone of anticlinal folding, with the center lying along fractures associated with a surface of a branch fault, relatively close to the main fault line.

In general a gryphon formation area A_g, can be divided into:

$$A_g = A_{g1} + A_{g2} \qquad (4.2)$$

where A_{g1} is the gryphon formation area on volcanoes, bounded by the mud-volcano crater area; A_{g2} is the area within which gryphons are scattered beyond the crater zone, usually along a fault line.

A statistical analysis of A_{g1} has enabled the distribution of volcanic crater diameters to be determined. Because the majority of craters are circular, the crater area is

$$A_{g1} = \pi D^2/4 \qquad (4.3)$$

where D is the crater diameter. Figures 4.1 and 4.2 show histograms of the distributions of these values, corresponding to an exponential distribution, with an average of 242m^2 which can be represented in terms of the length, L_f, of the fracture along which gryphons occur in a zone of width d, as $A_{g2} = L_f d$.

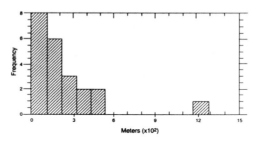

Figure 4.1. Frequency histogram for the crater diameters of mud volcanoes.

Unfortunately, the data compilations of Yakubov et al. (1974) and Rakhmanov (1987) do not provide sufficient data for evaluating the above parameters. Only in a few instances were data on L_f available, but the data were so sparse as to be unsuitable for statistical analysis. Similarly, data on the width d were very sparse. Therefore, it is necessary to undertake additional studies, including field work, in order to obtain the distribution of A_{g2}.

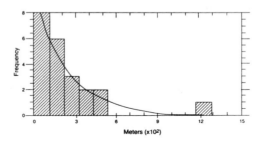

Figure 4.2. Exponential curve for the values of crater diameters of mud volcanoes.

5. Waiting time Predictions and Mud Volcano Frequency of Eruptions

Let the probability that there is an eruption of a volcano in Azerbaijan be p. Then the probability, P_k, that k eruptions out of n occur during a single year can be described by a Poisson distribution (Feller, 1971):

$$P_k = \exp(-\lambda)\, \lambda^k/k! \tag{4.4}$$

where $\lambda = np$ is the expectation of the number of eruptions in a year. Consequently the probability of k eruptions over t years is

$$P_k = \exp(-\lambda t) \, (\lambda t)^k / k! \tag{4.5}$$

Now the number of eruptions in all the volcanoes is the sum of the eruptions occurring for each volcano. Thus, the number of eruptions in Azerbaijan is equal to the sum of r independent random variables, describing the number of eruptions of r different volcanoes. But, if a random variable is the sum of independent random variables, each of which has a Poisson distribution, then the expectation of the total distribution will be equal to the sum of the expectations of each of the components. Accordingly, the problem reduces to that of evaluating the actual value of the parameter λ, while allowing for gaps in the record of eruptions. Two mechanisms exist by which gaps can arise. First a gap can be due to a weak eruption which is not observed; it is then necessary to provide a measure of eruptive power. If it is further taken that more weak eruptions occur than do powerful ones, and that the number of eruptions decreases in proportion to their intensity, one can evaluate the probability of weak eruptions.

Second, gaps can also arise in time. The further in the past one goes the less likely is the chance that an eruption was recorded. Indeed, only a small number of eruptions have been recorded at the beginning of the last century whereas, at the present, only a few weak eruptions are likely to go unrecorded. Accordingly, an accurate application of statistics requires that the period over which mud volcanoes are studied be broken down into intervals over which the average number of recorded eruptions remains fairly constant. It is then possible to find the probability of a gap in the eruption records by evaluating the time interval of observation.

It would, then, seem possible to evaluate the average number of eruptions which actually occurred in Azerbaijan (and may not be recorded), and then to use this information to determine the expected average number of eruptions per year in the Chirag area.

a. Determination of Eruptive Intensity.

Rakhmanov (1987) divides eruptions into four types:

Type I: explosions of varying strength, the emission of a large volume of gas with ignition, and the eruption of a large volume of breccia;

Type II: the eruption of breccia without ignition of gas;

Type III: emission of low-viscosity breccia with no intense gas emissions;

Type IV: extrusion of breccia and the emission of a negligible amount of gas.

For each eruption it is possible that mud may be emitted with an insignificant volume of gas, or vice versa. The absence of a strong correlation between these parameters will be discussed later.

In this chapter interest is also centered on the parameters of mud volcano eruptions on which would depend the probability of recording the eruptions; i.e. those parameters which are significant in determining the probability of a gap in the record. Accordingly, consideration has been given to parameters related to the erupted material (length, width, area, thickness, and volume of breccia), the height of the erupted breccia, the height of the flame and the combustion period, the duration of the eruption, the noise level and the occurrence of earth tremors. The greater the values of each of these parameters, the less likely is it that an eruption will be missed in the record. However, these parameters do not correlate with each other. Furthermore, gaps occur for all eruptions; i.e. not all the parameters are measured in every eruption. The problem is how to determine some measure of the intensity of an eruption which allows for each factor. To this end, cluster analysis was undertaken using the dynamic condensation method allowing for gaps in the data, based on end-member considerations (Bagirov and Nadirov, 1995). Eruptions of Table 4.4 were used as standard measures of grade intensity. Grade 1: Similar to eruption No. 233 with a parameter vector: (5, 5, 0.025, 0.5, 0.0125, 0, 0, 0, 0.1, 0); Grade 2: Similar to eruption No. 34 with a parameter vector: (35, 35, 0.8, 0.4, 0.3, 10, 0, 0, 0.5, 1); Grade 3: Similar to eruption No. 25 with a parameter vector: (100, 100, 8, 1.5, 10, 75, 200, 1, 1, 3); Grade 4: Similar to eruption No. 224 with a parameter vector: (230, 170, 30, 5, 160, 50, 400, 3, 1, 3); Grade 5: The post powerful eruption, similar to the eruption of Airantekyan, with a parameter vector: (1800, 50, 175, 4, 700, 400, 1000, 5, 5, 3); where the parameter vectors for each entry are from entries 4 through 14 in Table 4.4 in sequence. All eruptions can be compared to these standards. The class number in which the i-th eruption falls is called intensity. Therefore, five intensity levels are defined, which is the optimum number due to the low volume of records and the large number of gaps. The cluster analysis results are given in Table 4.4 where, in addition to the parameters used in the analysis, the last column provides the intensity of eruptions in more conventional units.

b. Variability of Observed Eruptions with time

Tracing the history of recorded Azeri mud volcano eruptions leads to the identification of irregularities in the recorded distribution, due to the increasing number of workers with time devoted to the study of mud volcanoes. During the last century eruptions were recorded only by chance and only by a few scientists (Abikh, 1939); however, during the second half of the present century these problems have occupied large groups of research geologists. Because there are no permanent stations for observing mud volcanoes, and because field work is undertaken sporadically, there still exists the possibility of failing to record a mud volcano eruption, especially a weak one. But this possibility is small compared to results reported 100-150 years ago. The problem here is to trace the recorded eruptions through time, and to use this information to evaluate any change through time in the probability of gaps occurring in the

68

record. Divide the entire time period over which observations have been made into intervals, within which the expectation of an eruption would remain

Table 4.4. Parameters of Volcanic Eruptions

1	2	3	4	5	6	7	8	9	10	11	12	13	14
3	95.8	22.6	300	200	60	2	120	-1	600	3	10	3	4
6	77.8	11	-1	-1	-1	-1	-1	-1	60	1	-1	1	3
8	69.9	4.9	-1	-1	5	-1	-1	-1	40	-1		-1	2
9	66.3	3.6	153	80	12	1	12	25	50	1	0.15	-1	1
11	57.8	2	250	100	25	-1	-1	-1	-1	-1	7	0	5
13	51.7	3.1	120	30	3.6	-1	-1	-1	-1	-1	-1	3	3
14	38.4	13.3	-1	-1	41.6	2.5	104	25	450	0.25	2	2	2
15	33	5.4	-1	-1	100	2	175	45	200	0.16	2	3	4
16	20.1	12.9	150	50	8	-1	-1	-1	250	0.05	-1	4	2
17	15.1	5	-1	-1	55	4	200	-1	200	2	2	-1	1
18	12.7	2.4	210	100	25	-1	-1	-1	0	0	-1	-1	4
24	65.5	25.6	110	110	10	2	20	-1	250	-1	-1	3	3
25	63.2	2.3	100	100	8	1.5	12	75	250	-1	-1	3	3
27	46.6	7.3	200	160	25	-1	-1	100	-1	-1	-1	-1	4
29	38.1	5.7	80	80	5	1.2	7	-1	150	-1	-1	1	2
31	18.6	5	115	100	9	1.8	16	-1	100	-1	-1	-1	1
32	12.1	6.5	100	100	8	1.5	12	-1	100	2	-1	3	3
34	1.7	4	30	35	0.8	0.4	0.3	-1	0	0	-1	-1	4
35	0.2	1.5	25	25	0.5	0.5	0.25	-1	0	0	1	1	4
41	40	37	-1	-1	300	1.25	500	35	-1	-1	0.2	2	2
44	26	2	190	-1	-1	-1	-1	-1	-1	-1	0.1	-1	2
45	24	2	285	55	18.7	0.8	15	25	-1	-1	0.12	-1	4
46	3	21	170	-1	-1	-1	-1	10	-1	-1	-1	1	4
48	75	-1	-1	-1	-1	-1	-1	-1	200	2	-1	-1	3
50	51	17	-1	-1	-1	-1	-1	-1	200	-1	-1	-1	3
51	40	11	-1	-1	-1	-1	-1	-1	200	-1	-1	-1	3
57	27	4	-1	-1	63	1.2	5	15	-1	2	-1	1	4
58	6.6	20.4	120	115	10	-1	-1	100	-1	1	-1	1	3
59	0.1	6.5	80	60	4.8	0.5	2.4	100	-1	1	-1	-1	3
60	116	-1	-1	-1	-1	-1	944	-1	-1	1	-1	-1	1
64	71	4	-1	-1	-1	-1	927	-1	-1	1	-1	-1	1
65	67	4	-1	-1	-1	-1	1500	20	0	0	-1	-1	4
68	34	17	-1	-1	-1	-1	-1	-1	200	0.5	0.03	3	3
77	25.9	40.4	-1	-1	200	2	400	-1	150	-1	-1	2	3
78	0.3	25.6	-1	-1	80	2	160	-1	500	-1	2	3	2
79	135.6	-1	359	-1	-1	-1	16	-1	-1	0.75	2	3	3
80	132.3	2.3	-1	-1	-1	-1	17.4	-1	-1	-1	-1	3	3
81	52.3	81	-1	-1	1300	-1	2000	-1	1000	1	2	-1	1
83	33.2	12.1	150	100	15	0.3	4.5	-1	-1	-1	-1	-1	3
84	32.5	0.7	500	450	125	1.5	270	200	-1	-1	0.15	3	3
87	55.6	66	-1	-1	2.8	-1	18	-1	-1	-1	0.01	-1	5
88	55	0.5	-1	-1	-1	-1	350	-1	-1	-1	-1	-1	2
89	54	1	-1	-1	65	-1	240	-1	-1	-1	0.0005	1	4
90	43	11	-1	50	-1	-1	150	-1	0	-1	0.02	-1	4
91	32	11	-1	-1	-1	-1	1160	-1	100	0.25	-1	-1	3
92	23	9	-1	-1	-1	-1	240	5	200	0.3	-1	-1	4
95	68.5	23	530	235	-1	-1	-1	-1	-1	-1	-1	-1	2
97	45	15.4	120	100	12	4	40	500	500	-1	-1	-1	1
99	7	35	140	140	5	0.5	8	-1	-1	-1	-1	-1	3
104	10	10	200	150	30	1.5	45	-1	6	-1	-1	-1	4
105	8	2	200	180	36	1	36	-1	-1	-1	-1	-1	3
106	6	2	300	200	60	1.2	72	-1	-1	-1	-1	-1	3
107	109.2	-1	-1	-1	-1	-1	25	-1	-1	-1	-1	-1	3
109	87.2	22	-1	-1	-1	-1	17	20	-1	-1	-1	-1	4
110	42.4	36.8	-1	-1	-1	-1	30	-1	-1	-1	-1	-1	3
111	39.8	2.6	-1	-1	-1	-1	50	30	-1	-1	-1	-1	2
112	39.4	0.4	-1	-1	-1	-1	35	15	-1	-1	-1	-1	4
113	139	-1	60	65	3	2	6	5	-1	-1	0.125	3	2
114	88.1	50.9	-1	-1	2500	-1	-1	-1	-1	-1	-1	3	1
115	71	17.1	-1	-1	1500	-1	-1	-1	-1	-1	-1	-1	1
116	41	30	300	230	70	3	210	-1	150	-1	-1	-1	3
117	27	14	230	230	50	1.5	170	-1	100	2	-1	-1	4
119	155	-1	600	400	240	1	240	-1	100	3	1	-1	3
122	39	51	-1	-1	0.04	-1	-1	-1	10	96	-1	1	1
123	35	4	150	6	15	1	9	-1	-1	-1	-1	-1	4
124	18	16.8	300	200	60	5	310	450	30	1	-1	3	1
125	5	13	60	60	3	-1	-1	-1	300	-1	-1	1	4
131	28	-1	2000	250	175	4	700	-1	150	3	15	3	1
132	23	5	-1	-1	450	-1	100	-1	40	-1	-1	-1	4
133	15	8	-1	-1	63	4	250	-1	1000	5	-1	3	1

Table 4.4 (cont.)

1	2	3	4	5	6	7	8	9	10	11	12	13	14
134	4	11	-1	-1	-1	-1	-1	-1	50	-1	-1	-1	4
135	2	2	-1	-1	-1	-1	-1	-1	120	-1	-1	3	3
136	44	-1	-1	-1	-1	-1	-1	30	0	-1	-1	3	3
138	10.1	13	220	220	38	1	38	-1	-1	-1	-1	-1	3
139	10	0.1	130	130	13	-1	-1	-1	-1	-1	-1	-1	3
140	26	-1	350	65	22	3.5	80	0	0	0	-1	0	5
142	15	7	300	20	8	-1	-1	0	0	0	-1	0	5
145	143	50	-1	-1	-1	-1	-1	-1	110	0.5	1	3	3
147	77	13	-1	-1	2000	-1	-1	-1	100	-1	-1	-1	3
148	56	21	-1	-1	-1	6	-1	-1	1000	0.25	-1	-1	1
158	66	4	-1	-1	-1	-1	18	-1	300	0.2	-1	-1	3
159	61	5	-1	-1	-1	-1	617	-1	150	9	-1	3	2
173	5	86	100	100	8	1	8	-1	0	0	-1	-1	4
174	53	-1	-1	-1	-1	-1	-1	20	300	30	-1	3	2
176	17	28	-1	-1	50	3	150	25	200	-1	-1	3	2
178	20	66	200	200	30	0.4	12	-1	0	0	-1	0	5
183	57	-1	500	200	65	2	125	-1	-1	-1	-1	-1	2
184	17	40	400	50	20	-1	-1	-1	-1	-1	-1	-1	2
188	22	13	120	100	10	-1	-1	-1	-1	-1	-1	-1	3
190	42	23	270	220	60	1.25	75	-1	20	0.5	-1	1	4
191	25	-1	-1	-1	30	1.5	35	150	0	0	0.05	2	3
192	15	10	90	90	6	-1	-1	25	0	0	0.005	3	3
193	46	-1	400	400	160	1	160	-1	-1	-1	-1	3	2
194	9	37	140	140	15	1	16	-1	-1	-1	-1	-1	3
195	31	-1	180	160	29	0.75	160	-1	400	3	-1	1	2
196	24	7	500	350	70	7	500	200	350	1.5	-1	3	3
200	22	100	60	50	3	6	20	0	0	0	-1	0	5
201	39	100	2300	1500	1000	3	2000	300	600	-1	0.05	3	1
202	17	22	-1	-1	-1	-1	20	-1	300	-1	-1	2	3
203	4	-1	150	150	18	0.5	9	-1	-1	-1	-1	-1	3
204	3	1	100	100	10	0.5	5	-1	-1	-1	-1	-1	3
211	67	-1	80	80	5	0.7	3	-1	-1	-1	-1	1.2	3
212	44	-1	80	70	5.6	1.3	7	-1	0	0	-1	-1	4
213	32	-1	700	300	210	3	630	-1	100	-1	-1	-1	2
221	43	-1	120	-1	-1	1.5	100	75	200	1.5	0.3	3	3
222	3	-1	100	100	8	1	8	-1	-1	-1	-1	3	3
223	2	-1	100	100	8	3	24	-1	500	-1	-1	3	2
224	30	-1	230	170	30	5	160	-1	400	3	-1	3	2
228	3	-1	70	70	4	0.5	2	-1	-1	-1	-1	-1	4
230	18	-1	200	200	31	-1	-1	25	0	0	0.1	1	4
233	11	-1	5	5	0.025	-1	-1	0	0	0	-1	-1	5
235	4	-1	80	80	5	0.5	2.5	-1	-1	-1	-1	-1	4
238	32	-1	-1	-1	200	10	2000	5	250	0.7	-1	3	2

Value of parameter equal to -1 signifies that it was not recorded.

Key:
1. Sequence number of eruptions.
2. How many years ago the eruption occurred.
3. Time since the last eruption of this volcano (years).
4. Length of tongue of erupting breccia (m).
5. Width of tongue (m).
6. Area covered (thousand m^2).
7. Thickness of cover (m).
8. Volume of breccia erupted (thousand m^3).
9. Height of eruption (m).
10. Height of flame.
11. Combustion time (hours).
12. Duration of eruption (days).
13. Noise (1: rumble; 2: rumble with tremors; 3: explosive bang).
14. Conditional strength of eruption (in balls) from the results of cluster analysis

70

constant. Then a cumulative time series approach can be adopted; i.e. assignment made of a value equal to the number of eruptions which had occurred up to a particular year. The curve of the cumulative time series is shown in Fig. 4.3. The problem was then reduced that of finding the points where the expectation changed abruptly ("step-points"). The entire time interval was divided into seven intervals (Fig. 4.4): (i) 1810-1824; (ii) 1825-1853; (iii) 1854-1900; (iv) 1901-1911; (v) 1912-1940; (vi) 1941-1986; (vii) 1987-1992.

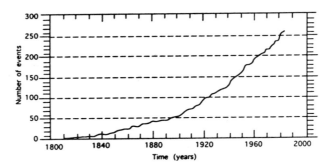

Figure 4.3. Cumulative time sequence of the number of eruptions per year.

Two eruptions occurred during the first time interval; 9 in the second; 23 in the third; 13 in the fourth; 56 in the fifth; 109 in the sixth; and 227 in the seventh. The

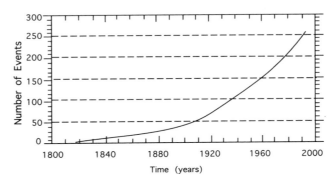

Figure 4.4. Results of smoothing the time intervals of observations.

λ parameter for each interval, which we designate by a subscript, is (in units of yr^{-1}): $\lambda_1 = 0.13$; $\lambda_2 = 0.41$; $\lambda_3 = 0.6$; $\lambda_4 = 1.18$; $\lambda_5 = 1.93$; $\lambda_6 = 2.28$; $\lambda_7 = 4.67$. Fig. 4.5 provides a graph of the changes in the observed average number of eruptions per year.

Consider now the distribution of eruption intensity measured in conventional units (see Table 4.4). Recall that in a number of eruptions not a single parameter was recorded, and hence it is not possible to assess the intensity of those eruptions. Thus the distribution shown in Table 4.5 is for observed eruption intensities.

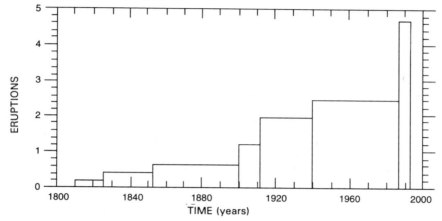

Figure 4.5. Average values of the number of recorded eruptions per year in Azerbaijan.

Table 4.5. Intensity of Eruptions over Different Periods of Time

Period	Observed Intensity of Eruptions				
	1	2	3	4	5
1810-1824	0	0	0	0	0
1825-1853	0	0	1	0	0
1854-1900	1	1	4	1	0
1901-1911	1	0	0	0	0
1912-1940	5	6	7	4	2
1941-1986	8	12	26	15	5
1987-1992	0	2	5	7	0
TOTAL	15	21	43	27	7

Despite the sparseness of each grouping, the observed distribution of intensities does not change within statistical significance in the different time periods; homogeneity criteria (such as martingale statistics) are not effective for negating this hypothesis due to the sparse data. Accordingly, consider the two event classes: "gap, depending on remoteness in time" and "gap, depending on intensity of eruption" as independent, and study separately the probabilities of these events.

c. Evaluation of the average number of eruptions per year

Each eruption may be either recorded or missing, with a probability which depends on the intensity of the eruption itself and on the time when the eruption occurred. Introduce a gap indicator as: $R_i=0$ if the i-th eruption was missed, and $R_i=1$ if observed. Then: with $P\{R_i|J_i,T_i\}$ as the conditional gap distribution with a specific strength J_i and time of eruption T_i; $P\{R_i|J_i\}$ as the conditional gap distribution with specific strength, J_i; and $P\{R_i|T_i\}$ as the conditional distribution of gaps with a defined time, T_i, one has:

$$P\{R_i|J_i,T_i\} = P\{R_i|J_i\}P\{R_i|T_i\} \qquad (4.6)$$

These probabilities are now evaluated. Take the probability of a gap in the eruptions, falling in the j-th interval, as:

$$P(R_i = 0|T_j) = 1 - \lambda_j/\lambda_7 \qquad , \qquad j=1,2,...6 \quad ; \qquad (4.7a)$$

and

$$P\{R_i = 1|T_j\} = \lambda_j/\lambda_7 \qquad (4.7b)$$

The evaluation of $P\{R_i|J_i\}$ is more complex. Make the additional assumptions that there are more weak eruptions in nature than strong ones, and that grade four and grade five eruptions will not be missed. Then

$$P\{R_i=1|J_i=5\} = P\{R_i=1|J_i=4\} = 1$$

Construct a graph of the recurrence of eruptions in order to evaluate the actual number of eruptions of grades 1, 2 and 3 (Fig. 4.6). Let n_k be the number of recorded eruptions of grade k, and n_k^* the actual number of eruptions of grade k, obtained by extrapolating the graph of figure 4.6 for k=1, 2 and 3. The results are summarized in Table 4.6.

Table 4.6. Recorded and Actual Number of Eruptions for Grades 1,2 and 3

Grade Value	Number Recorded	Actual Number	Probability of Occurrence
k	n_k	n_k^*	$P\{R_i=1|J_i=k\}=n_k/n_k^*$
1	15	87	0.17
2	21	67	0.31
3	43	47	0.92

The total number of eruptions for which it was possible to define an intensity was 113. Accordingly the actual number of eruptions, N, estimated to have occurred in Azerbaijan over the last 183 years is:

$$N=\lambda_7(2/\lambda_1+9/\lambda_2+23/\lambda_3+13/\lambda_4+56/\lambda_5+109/\lambda_6+27) \, (n_1+n_2+n_3+n_4+n_5)/113 = 1.8 \times 10^3,$$

corresponding to an average of about 10 eruptions per year, the majority of which have been weak (grade 3 or less). Because eruptions of grades 1 and 2 make up 154/235 of all eruptions, the total number of eruptions, of grade 3 or higher, is estimated at 625 for an average of 3.5 eruptions per year.

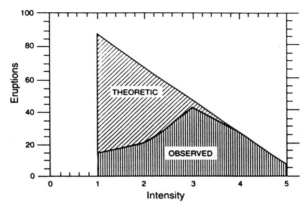

Figure 4.6.   The recurrence of eruptions with superposed theoretical predictions.

d. Expected number of yearly eruptions at the Chirag area

On average the total number of eruptions of mud volcanoes in Azerbaijan is 10/yr and, if weak eruptions are ignored, then this number is about 3.5/yr. But these eruptions are not uniformly distributed across all volcanoes. Indeed, 19 eruptions have been recorded for the Lokbatan volcano, and 16 for Shikhzagirli, whereas the large majority of volcanoes have one or zero recorded eruptions. Table 4.7 provides the frequency of recorded volcanic eruptions in Azerbaijan.

Table 4.7. Recorded Volcanic Eruptions

Number of eruptions	1	2	3	4	5	6	7	8	11	12	16	19
Number of volcanoes	33	13	6	6	3	2	5	1	2	1	1	1

It would seem appropriate to allocate the number of eruptions per year (λ) to each volcano in proportion to the recorded frequency, so that the sum of the values of λ for all the volcanoes on which eruptions have been recorded will be equal to 10 and 3.5, respectively, for the total average number of eruptions per year, and for the average of strong and medium eruptions per year. Simple calculations show that volcanoes on which k eruptions have been recorded may be assigned an average number of all eruptions per year, λ_{1k}, and an average number of strong and medium eruptions of λ_{2k} per year, as given in Table 4.8.

To evaluate λ_{10} and λ_{20} from the data in Table 4.8, i.e. the average frequency of eruptions for those volcanoes in which not a single eruption has been recorded, it is first necessary to select a function describing the relationship between λ_{1k}, λ_{2k} and the frequency, v_k. Now 240 eruptions are recorded from 74 volcanoes, and 163 of the eruptions are recorded from just 22 volcanoes on which there have been four or more eruptions. Therefore, approximately 70% of eruptions were recorded from just 30% of volcanoes, suggesting that a Pareto-type of distribution (Feller, 1971) is likely appropriate. Thus one writes

$$v_k = a_1(\lambda_{1k})^{-b_1} \tag{4.8}$$

or

$$v_k = a_2(\lambda_{2k})^{-b_2}$$

where $b_2 = b_1$ because $\lambda_{2k} = (3.5/10)\lambda_{1k}$, by definition.

Regression analysis with the data of Table 4.8 provides the values

$$a_1 = \exp(-0.59); \quad a_2 = \exp(-1.88); \quad \text{and} \quad b_1 = -1.23 \equiv b_2,$$

with correlation coefficient 0.94. Extrapolating to the domains $\lambda_{1k} < 0.0416$ and $\lambda_{2k} < 0.0146$ (corresponding to k < 1) yields the frequency $v = 150$, corresponding to the values $\lambda_{10} = 0.0105$, and $\lambda_{20} = 0.00368$, which apply to the average number of eruptions per year for volcanoes on which not a single eruption has been recorded.

The probability distribution, p(t), of the waiting time over a time interval t (yrs), for the first eruption of <u>any</u> size on the Chirag mud volcano then takes the form:

$$p(t) = 0.0105 \exp(-0.0105t) \tag{4.9a}$$

and for the first eruption of strength grade 3 and above, the corresponding form is

$$p_3(t) = 0.00368 \exp\{-0.00368t\}. \tag{4.9b}$$

The corresponding expectations of the waiting time are 95 and 272 years, respectively. The number of expected eruptions over t years is 0.0105t for all eruptions, and 0.00368t for strong and medium eruptions.

Table 4.8. Average Values of Yearly Eruptions for Different Volcanoes

Number of Yearly Eruptions k	Yearly Average (All) λ_{1k}	Yearly Average (Medium & Strong) λ_{2k}	Frequency of Occurrence, ν_k
1	0.0416	0.0146	33
2	0.0832	0.0292	13
3	0.1248	0.0438	6
4	0.1664	0.0584	6
5	0.208	0.073	3
6	0.2496	0.0876	2
7	0.2912	0.1022	5
8	0.3328	0.1168	1
11	0.4576	0.1605	2
12	0.4992	0.1752	1
16	0.6656	0.2336	1
19	0.7904	0.2774	1

The probability that not a single eruption of grade 3 or above will take place on the Chirag volcano over 100 years is then only 30%, i.e. there is a 70% chance that an eruption of grade 3 or above <u>will</u> occur. Figures 4.7 and 4.8 show the cumulative probability functions for the waiting time of all eruptions, and for grade 3 and above eruptions, respectively.

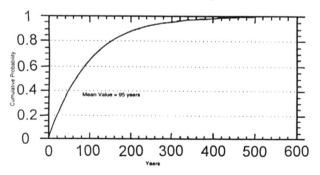

Figure 4.7. The cumulative probability of the waiting time for eruptions on the Chirag volcano.

B. Flaming Eruptions. Historical Review

One of the hazards presented by mud volcanoes is flame development. To achieve a better understanding of the spatial scales and actual sizes of flames, which can destroy equipment and people, this historical overview has been

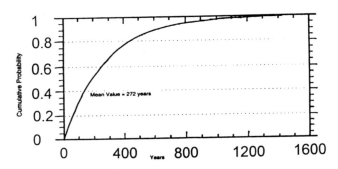

Figure 4.8. Cumulative probability of the waiting time for strong and average strength eruptions on the Chirag volcano.

developed. All case descriptions were taken from published papers of Azeri geologists, available mostly in Russian. Here are some of them.

On September 4, 1961 a strong eruption occurred on Duvanny Island (located 12 km offshore and 37 km to the south-west of Baku), which lasted 45 minutes. According to Sultanov and Dadashev (1962) and Yakubov et al. (1970a) the eruption started with a muffled underground rumble. Ground slowly lifted 10-15 meters. Then black smoke appeared, which was followed by a flame of 200-250 meters height. The temperature around the eruptive center increased and wooden constructions, at a distance of 300 meters away, inflamed. On the upper part of the emission a mushroom shape cloud formed, which started to condense in 5-10 minutes and the temperature decreased. The flame, however, burned for 30-45 minutes.

The diameter of the erupted mud cover was 400-600 m and the thickness was 5-7 meters. The mud was almost dry. Spherical empty rocks ("lapilli") were found on the coast about 30 km away from the island.

Another offshore eruption, which happened on Makarov Bank, was described by Sultanov and Agabekov (1959) and Yakubov et al. (1970a). The Makarov Bank is located within the offshore Bahar oil-gas field, about 50 km from Baku. A very strong eruption occurred October 15, 1958 (at 21.50). Explosion was followed by a flame of 200 m height and 50 m in diameter. The color of the flame was red-orange, was very bright and was observed from Baku and Oil Rocks Field, which were 80 km away. The flame was watched from Akhsu Pass (about 200 km away).

The captain of the ship Fataliyev watched the fire from the beginning. About 10 p.m. he was at a distance of 27 km from the eruption center. But it seemed that the fire was very close and they changed the direction of the ship to the south. A huge surface area of the water was on fire. Some flashes of flame moved mostly in the direction of the wind, which blew that day from north-east to south-west. Then the ship continued its way to Baku Bay. When they were about 18 km from the eruption center the crew felt that the deck was covered by sand. And 30-35 minutes after beginning the flame was extinguished. The next day the crew found all the deck covered by black sand. The density of the cover was approximately 1 handful from 1 m^2. The grain sizes were from decimals of millimeters to 2 millimeters. The grains crunched under foot pressure because they were hollow, like cored spheres. That probably happened when the sand particles were sucked into the flame. Under high temperature sand particles blew out due to expansion of inner moisture, and sometimes burst. There are small bubbles on the surfaces of the sandstone particles. Such particles probably are erupted on other volcanoes as well. But, because they are very small, they went unnoticed.

The interesting thing was that the wind blew from north-east to south-west (most likely) and the vessel was to the north-east from Makarov Bank at that time (that fact was mentioned a second time), and was at a distance of 18 km from the eruption center. Sultanov and Agabekov (1959) explain the observed phenomena by the fact that wind can have different directions at different heights. While the air strata near the surface had a NE-SW direction, the strata above (at heights of 1000-1200 m) can have an oppositely directed flow. So the small particles could be lifted by the flame column to higher levels and then transported over far distances. According to Yakubov et al. (1970a) 2-3 explosions were recorded during the eruption.

Eyewitness accounts and newspaper records of other flaming eruptions also exist. Thus, for the Ayrantekyan mud volcano, located onshore at 82 km to the south-west of Baku, it is noted that an eruption occurred on October 6, 1964 about midnight, and was observed from both the drilling rig and the small village of Khanaly, located on the south-west slope of the volcano (Yakubov et al., 1965 and 1970a).

During the night of 6 to 7 October 1964 Ayrantekyan volcano erupted. An eruption was observed from the drilling rig, approximately 1 km away from the eruption center. A muffled rumble was followed by explosion and a flame of height 100-150 m arose over the volcano. The area around the volcano was illuminated to a distance of 10-15 km. The flame was wider in the upper part and had a shape of a turned-over cone. The fire burned 20-30 minutes and then got weaker. Its height decreased to 20-100 m and the shape had more regular contours. The temperature at the rig increased and it was hard to breathe. At the beginning the flame was whitish and later black smoke appeared. When mud started being thrown, the intensity of the fire decreased; 4-5 hours after the eruption its height was 10-20 m and then later 5-2 meters. The eruption center crater had a diameter of 250-450 m. To October 8, the height of flame remained 1.5-2 m, but even on October 21, the gas still burned (0.5 m height).

The ground to the north of the crater (500-600 m away) was burned and had a dark-brown color, while the surface of the fresh mud was reddish-brown. Bushes were burned from the flame side of the volcano. According to the melted breccia, and comparing with lab analysis, the temperature of burning gas was 1200-1400°C. Burning gas seeped from different fractures as well.

The land surface probably burned at the first moment of the flame, because at the edges of the mud cover one could see later fresh flows covering parts of the burnt surface.

The next observed eruption on Ayrantekyan occurred on 10 June 1969 (5 p.m.) (Yakubov et al., 1970a). During the first few hours the height of the flame reached 400 m. By the morning of 11 June the height of the flame was down to about 15 m, but a very strong wind was bending the flame. By noon scientists from the Geological Institute of Azerbaijan approached the crater. By that time the flame was only 1.5-2 m high but, because of the wind, the flame was bent and from time to time licked across the ground. A week later the gas was still burning and the height of the flame was about 1 m. The temperature in the crater during eruption was estimated to have reached 800-1000°C.

Very interesting observations of the Bolshoi Kanizadag eruption were made by Gorin on May 12, 1950. This volcano is the biggest for onshore Azerbaijan and, during the prior 150 years, there was no activity of this volcano. We present here the translation of Gorin's paper (with elimination of some details) (Gorin, 1950):

"I watched this eruption from the very beginning and then compared my observation with those of other geologists, who also watched the eruption.

The eruption started about 6 a.m. An explosion first and then a flame appeared at 6:15 a.m. However, most likely, breccia ejection happened before that. Because when I reached the edge of the mud flow at 8:15 a.m. there was already a dry crust on it and the velocity of the mud flow was very slow, which shows that the mud ejection happened some time at night.

The first stage of the fire lasted from 6:15 a.m. to 7:51 a.m. There was enough time to draw a picture (The picture in Gorin's paper shows that the height (not the length) of the flame is approximately one third the height of the 400-450 m high volcano). There was a strong force 4-5 north wind and the angle of the flame to the horizon reached 45° and the height 150-200 meters. The picture was drawn from a point 4-5 km to the south-east of the volcano. The flame column consisted of the shooting up every second of reddish-white puffs of fire, which rotated like fire balls. The smoke had an ash-grey color. At 7:51 a.m. the fire and smoke stopped. The rumble stopped as well. Such a temporary silence was maintained during all the time that I climbed up the hill.

At 8:30 a.m. I stayed at the edge of the mud-covered area, about 200 meters from the crater. Every minute pieces of mud were thrown to a height of 10-15 meters. Going around I saw spots of completely burned grass and wormwood, surrounded by areas covered with grass and not affected by heat action. Such burned areas were mostly on the south and west sides from the crater. It seemed that the rapidly flying fire balls just touch the ground in some places. From the leeward side of the crater it smelled of sulfur and burned grass.

At 8:47 a.m. gas started to erupt again. It came out with low level noise and ignited at a height of approximately 5 meters and then formed those fire puff-balls. Those balls came periodically, like pulsations. At the same time the pieces of thick mud were thrown to a height of 5-10 meters. Such conditions kept going for 12 minutes until 8:59 a.m. This phase was quiet and allowed one to watch an eruption from a distance of 300 meters. The burning gas puffs reached diameters of 10 meters at a height of 15-20 meters and then got smaller and disappeared. From 8:59 to 9:31 a.m. the gas eruption stopped. In 32 minutes the gas emission and ignition started again. But 6 minutes later a strong pushing out of the mud column occurred. With a high rumble fragments of rock were thrown to a height of 70-80 meters during 15 minutes. Burning gas was not observed that time.

At 9:52 a.m. suddenly a silence came. It seemed that the crater was blocked up by the rocks, but from 10 a.m. mud ejection started again and stopped only at 12:30.

As one can judge, the eruption was characterized mostly by gas emission, rather than mud ejection. Mud cover does not exceed 1.5 m thickness and it did not cover even the whole top of the volcano."

A very strong eruption of Kelany volcano have been described by Yakubov et al. (1970b):

"December 12, 1969 the people of the village Polatly were witnesses of the eruption of Kelany volcano, taking place at 8:40 p.m. It started with a high rumble and tremors. Three to five minutes later a strong explosion happened and the flame of 350-400 meters height illuminated the area for 40-50 km around. About 10 minutes later a huge mass of mud erupted. The flame had an orange-red color with puffs of black smoke. Some rock fragments were thrown to a height of 150-200 meters and fell down not only on the erupted mud flow, but also far from that place (1 km away). Twenty minutes later the intensity of the volcano reduced. The height of the flame dropped by a factor two. But the volcano continued to throw rocks to a height of 45-60 meters and burning gas came from six different vents.

At 9:15 p.m. the rumble and the tremors got stronger, followed by a deafening explosion, and the flame column rose to a height of 250-300 m. The mass of erupted mud increased. Such explosions repeated at 10 p.m., 10:15, 11:10, 11:25 and 11:35 p.m. Very intensive burning lasted 5-15 minutes, then the intensity dropped. The last gas eruption (11:35) was the strongest. The flame reached 400 meters height.

By December 15 the flame height in the center of the crater was 8-10 m, and then reduced, and for a couple of days was 0.3-0.5 meters. The erupted mud covered 70,000 m^2 with an average thickness between 3-8 m. As a result of the fallen stones holes on the mud surface formed with diameters 3-5 meters and depths 1-2 m. In the center of the volcano burned places (10 m x 25 m) were found. There are a number of such places."

The Lokbatan mud volcano is the most active in Azerbaijan; some of its eruptions have been accompanied by flames. One such eruption was described by Zhemerev (1954):

"July 30, 1951 (22:22 p.m.) that was a strong eruption on Lokbatan. A crater was observed 2 days before an eruption, and no other evidence was noticed. We were at the foot of the volcano 1-2 minutes before the eruption. First an underground rumble similar to far peals of thunder and a hissing of gas were heard. Almost at the same time we heard a strong explosion. That was caused by the inflammation of gas. The flame column had a height of 400-500 meters and diameter 40-50 meters. The crater surface of 50 m x 100 m was on fire. The flame tongues reached 10-15 meters. The flame was visible in Baku, Alyat (about 70-80 km away). An area of 3-6 km in radius around the volcano was lit up as in daytime. Twelve minutes after the beginning of the eruption the flame dropped very fast and then gas burned on an area of 30 m x 70 m and formed separate tongues of 10-15 meters height. We approached the crater in 15 minutes. The grass around the crater at a distance of 100-120 meters continued burning. It was difficult to breathe, probably because of high concentration of CO_2 and CO. The temperature of erupted mud was then approximately 40-50°C.

The rock fragments of 2 to 5 kg mass and a lot of smaller fragments were distributed around a radius of 100 meters. The rocks falling to the ground cracked and then scattered to a distance of 1-1.5 meters, which indicates that they fell from a considerable height.

The first minutes after the eruption a cloud of dark-grey color arose over the flame to a height of 1.5-2 km. It was formed by steam, smoke and small clay particles. Those particles returned and formed a "clay rain", which started 20-30 minutes after the eruption started and lasted 1 hour and 30 minutes. At a distance of 3 km the volcanic dust condensed on the roads and crunched under the feet, which was noticed the next day. This "dust" consisted of baked clay particles of different shapes and 2 mm and less in size. The color was mostly light-grey and dark-brown.

All night and the next day gas burned, but the flame was weaker. The next day July 31 at 4:30 p.m. the volcano erupted again. This time an eruption occurred from the side crater. The flame had a height of 60-70 meters and burned for 2 minutes. Puffs of black smoke were over the flame. Particles of different sizes were pushed to a height equal to a third of the flame height. No mud was ejected this time. During the eruption fractures of 1 m width formed in an area of 400 meter width. The subsidence of the blocks in some places reached 2 meters. The subsidence was noticed on the morning of July 31 and this process finished August 3. The surface burning continued until August 11 and then it continued underground and in fractures. As a result the mud burned and had a brick color. At the beginning of the eruption blue-white smoke came from the fractures and smelled of chlorine and sulfur."

Another huge eruption of the Lokbatan volcano was observed in 1977 by researchers of the Geological Institute of Azerbaijan. This time the flame height reached about 400 meters.

Buniat-zade and Gorin (1968) described the eruption of Bahar volcano as follows:

"First started with tremor, then rumble and at 9 p.m. a huge orange-red flame of 100-200 meters height appeared. About 2 million m^3 mud was taken to the surface. Two days later there were two small flames of 2-2.5 meters height. Ten meters around the grass was burned. Empty spherical burned rocks of large pea-size were around."

Another strong eruption occurred on the same volcano in September 1992. The eruption ejected about 2×10^6 m^3 of mud, and the height of the flame reached about 400 meters. Ten days later a joint expedition of BP/STATOIL and the Geological Institute visited the volcano (one of the authors of this book (EB) was part of this expedition); a flame of about 1-1.5 meter height still burned.

Yakubov and Salayev (1953), and Nadirov and Zeinalov (1958) described two eruptions of Bozdag Kobiiski (August 1953 and August 1957) with a small (1 m) flame burning until the first rains. There are tens of other papers on flaming eruptions. But the length of this book precludes us from describing all of them here (but see Appendix C for a few descriptions).

C. Factors Associated with Flaming Eruptions

1. Composition of mud volcano gases

The chemical composition of gases emitted from mud volcanoes varies within fairly tight limits, and does not change significantly within the region. Gases emitted during eruptions have the same characteristics as gases emitted from gryphons and salses during non-eruptive phases. All the known Azerbaijan volcanoes are characterized by a high methane fraction (up to 99%), with carbon dioxide as the other major component. Ethane and heavier hydrocarbon gases are present only in minor amounts, as is nitrogen. Hydrogen and helium fractions are very low; of order 10^{-3} %.

Studies have shown that gases from mud volcanoes are almost devoid of hydrogen sulfide (10^{-5} %). This low sulfur property is also characteristic of oil and gas within economic accumulations, and of organic matter. Therefore, a low sulfur content in organic matter, and in oils and gases in the sedimentary sequence, is characteristic of the region.

Table 4.9 represents the chemical composition of gases from 72 different mud volcanoes in Azerbaijan (Dadashev, 1963; Valyaev et al., 1985), while Table 4.10 presents statistical characteristics for different components of the gases, showing directly the relatively low spread of values and a small mean-square deviation. A beta-distribution has been chosen to characterize the components because the fraction of each component varies from 0 to 1, and also because this distribution is flexible.

Table 4.9. Composition (%) of gases from mud volcanoes of Azerbaijan

CH_4	C_2H_6	Heavy HC	He	H_2	N_2	CO_2
97.11	0.09	0.12	0.0005	0	0.52	2.25
90.69	0.00	0.00	0	0	0.1	9.21
95.4	0.00	0.07	0.009	0	0.7	3.83
96.9	0.53	0.53	0.0058	0.01	0.96	1.6
98.53	0	0	0	0	0.75	0.71
97.54	0	0	0	0	1.06	1.39
95.42	0.08	0.09	0.0067	0	0.11	4.38
92.5	0.04	0.04	0	0	1.45	6.01
78.1	0.14	0.15	0	0	19.02	2.73
80.52	0.03	0.03	0.01	0	11.56	7.87
98.21	0.07	0.08	0	0	1.71	0
93.69	0.14	0.15	0	0	0.92	5.25
94.63	0.13	0.13	0	0	0	5.23
97.25	0.47	0.48	0.0013	0	0	2.27
97.55	0	0	0	0	1.04	1.42
98.11	0.04	0.04	0	0	0	1.85

Table 4.9 (cont.)

CH$_4$	C$_2$H$_6$	Heavy HC	He	H$_2$	N$_2$	CO$_2$
99.06	0.03	0.04	0.0018	0	0.1	0.91
96.71	0	0	0	0	0.74	2.55
94.57	0	0	0	0	0.22	5.21
96.23	0.03	0.04	0.001	0	0.15	3.58
98.16	0.02	0.03	0.0014	0	0	1.82
94.99	0.03	0.03	0.0014	0	0.69	4.3
97.1	0.03	0.03	0.0003	0	0.8	2.07
89.12	0.57	0.59	0.0011	0	0	10.29
90.75	0.03	0.04	0.018	0	0	9.21
99.47	0.01	0.02	0.0017	0	0.31	0.2
96.67	0.03	0.03	0.0033	0.0080	0	3.3
94.02	0.01	0.02	0	0	0	5.96
90.28	0.03	0.03	0.0012	0	0	9.69
93.82	0.03	0.03	0.0021	0	0.45	5.7
93.32	0.03	0.03	0.0026	0	0	6.65
95.1	0.03	0.03	0.0013	0	0	4.87
98.92	0.1	0.1	0.0014	0	0	0.99
99.54	0.04	0.04	0.0082	0.007	0	0.42
97.39	0.9	0.91	0.0033	0	0	1.7
93.39	0.03	0.03	0.0084	0	0	6.92
92.16	0.12	0.13	0	0	0	7.71
89.9	0.09	0.16	0	0	0	9.94
95.79	0.03	0.04	0.0006	0	0	4.17
96.05	0.4	0.43	0.0012	0	0	3.52
98.91	0	0.1	0.0006	0	0	0.99
97.73	0.03	0.03	0.0005	0	0	2.24
97.47	0.03	0.04	0	0	0.84	1.65
96.91	0.04	0.04	0.0006	0	0.01	3.04
98.58	0.01	0.01	0	0	0	1.61
99.03	0.03	0.04	0.0006	0	0	0.93
98.31	0.08	0.08	0	0	0.35	1.28
94.41	0.05	0.06	0.0025	0	0	5.53
98.88	0.05	0.05	0.0003	0	0	1.07
93.75	0.09	0.11	0.0005	0	0	6.14
99.13	0.12	0.12	0.0004	0	0.07	0.68
98.43	0.03	0.03	0.0007	0	0	1.54
98.74	0.06	0.06	0.0012	0	0	1.2
97.46	0.42	0.46	0.0034	0	1.54	0.54
97.63	0	0.01	0.0025	0	0.1	2.26
92.43	0.02	0.02	0	0	0	7.55
97.42	0.04	0.04	0.0015	0	0.48	2.06
98.78	0	0.01	0.006	0	0.23	0.98
94.06	0.1	0.1	0.0003	0	0.82	5.02

Table 4.9 (cont.)

CH$_4$	C$_2$H$_6$	Heavy HC	He	H$_2$	N$_2$	CO$_2$
96.06	0.02	0.02	0	0	3.27	0.65
98.56	0	0.01	0.0031	0	0.56	0.87
98.26	0	0.01	0.0024	0.05	0.87	0.81
98.99	0.06	0.06	0.0033	0	0.15	0.81
98.25	0.05	0.09	0	0	1.22	0.48
97.26	0.04	0.05	0	0	1.14	1.55
94.39	0.02	0.03	0	0	1.15	4.43
96.4	0.85	0.98	0	0	1.57	1.06
96.84	0	0	0	0	0.94	2.23
97.5	0.36	0.40	0.0032	0	0.81	1.29
93.45	0.71	0.73	0	0	4.43	1.4
94.25	0.00	0.48	0.0014	0	4.33	0.94
90.48	0.42	0.48	0.001	0	0.33	8.7

Table 4.10. Parameters for the β-distribution for Different Gas Components

	CH$_4$	C$_2$H$_6$	Heavy Hydrocarbs.	He	H$_2$	N$_2$	CO$_2$
α	26.6	0.15	0.17	0.36	0.029	0.11	1.34
β	1.2	91.11	62.93	19,727	2806	11.7	39.1

A random variable, x, has a beta-distribution, f(x), with parameters (α, β) where, in terms of the Γ-function, one can write:

$$f(x) = \Gamma(\alpha+\beta) \, x^{\alpha-1}(1-x)^{\beta-1}/\Gamma(\alpha)\,\Gamma(\beta), \text{ if } 0\leq x\leq 1; \, f(x)=0 \text{ if } x<0 \text{ or } x>1.$$

(4.10a)

The expected value of x^k is

$$E(x^k) = \Gamma(\alpha+k) \, \Gamma(\alpha+\beta)/\Gamma(\alpha)\Gamma(\alpha+\beta+k)$$

(4.10b)

while the variance of x, $\sigma(x)^2$, is:

$$\sigma(x)^2 = \alpha\beta/(\alpha+\beta)^2(\alpha+\beta+1)$$

(4.10c)

The parameters of the β-distribution for the different gas components, obtained by the maximum likelihood method, are presented in Table 4.10. The appropriateness of the beta-distribution was checked using the Kolmogorov-Smirnov and chi-square criteria, as shown on Figs. 4.9 through 4.13.

84

(Hydrogen and helium distributions were not constructed due to the low values). For the beta-distribution the expected value is $\alpha/(\alpha+\beta)$.

2. *Flame Ignition During Eruptions*

Not all eruptions are accompanied by flame, and flame ignition appears to depend more on the volume of breccia ejected rather than the volume of gas available for ignition. The cause(s) of flame ignition are still not clearly resolved. As reported in Abikh (1939) and Rakhmanov (1987): Gilev (1872) invoked an increase in temperature due to friction and the presence of sulfur to explain the ignition; while Gumbel (1879), as reported in Rakhmanov (1987), suggested the presence of hydrogen phosphorous. Archangel'skii (1883) demonstrated that iron sulfur compounds, occurring in large quantities in bituminous clays and shales, are readily oxidized by air-contact during an eruption. Sjögren (1888a,b; 1897) related an increase in the gas temperature to spontaneous combustion of methane (300-400°C) with friction and the outflow of deep gases; while Plotnikov (1896; as reported in Rakhmanov, 1987) considered the ignition to be due to a thermal chain reaction. Golubyatnikov

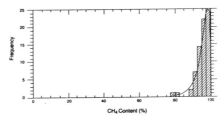

Figure 4.9. Histogram of the distribution of CH_4 content (%) in gas from mud volcanoes.

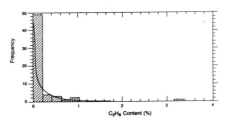

Figure 4.10. Histogram of the distribution of C_2H_6 content (%) in gas from mud volcanoes.

(1904) explained the presence of a flame by friction and the shocks sustained by the rocks acting on each other as they are ejected, and Shteber (1941; as

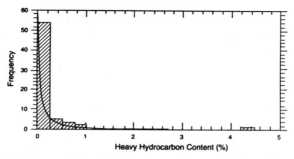

Figure 4.11. Histogram of the distribution of heavy hydrocarbon content (%) in gas from mud volcanoes.

reported in Rakhmanov, 1987) considered that a charge, built up as a result of collision between positive and negative electrical charges in the gas jet and the upper levels of the atmosphere (at a height of not less than 1500 m), created a spark discharge.

According to Rakhmanov (1987), spontaneous combustion is

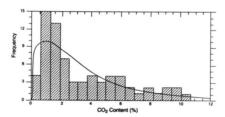

Figure 4.12. Histogram of the distribution of CO_2 content (%) in gas from mud volcanoes.

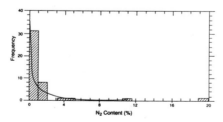

Figure 4.13. Histogram of the distribution of nitrogen content (%) in mud volcano gases.

associated with the gas composition and the presence of water-bearing horizons in the parts of the mud volcano close to its mouth. In the presence of a large volume of methane a chain reaction may begin, culminating in an explosion and, under favorable conditions, ignition of the gas mixture. Water acts as an active catalyst in this process, increasing the rate of the chain reaction by more than a factor of 1000. Due to the presence of large amounts of carbon dioxide in the mud volcano gases, or the absence of the water-bearing horizons, the rate of the chain reaction may reach a constant level at which an explosion occurs without flame combustion of the reacting gaseous mixture.

Ivanov and Guliev (1986, 1988) made the first attempt to create a quantitative physical-chemical model of mud volcanic processes. They emphasized that the actual pathways carrying the gas-laden mud are characterized by complex geometrical structures, including regions where the pathways narrow and widen. If the rate of flow in a narrow throat reaches the local speed of sound then, in a subsequent expansion, the gas is accelerated to supersonic speeds and, in a subsequent constriction, may slow to below the local speed of sound. In each case an abrupt density change occurs within the flow (pressure wave). The increase in temperature of the gas, ΔT, due to the pressure wave can be estimated from the classical formula for a DeLaval nozzle as:

$$\Delta T = 2T_0(k-1)\,(M_1{}^2-1)\,(1+kM_1{}^2)/(k+1)^2\,M_1{}^2 \qquad (4.11)$$

where ΔT is the temperature jump; T_0 is the original temperature of the gas before the pressure wave; k is the ratio of specific heats; and $M_1 = v/c$, the local Mach number, equal to the ratio between the rate of flow and the local speed of sound in the gas, c. Because the ignition temperature of a methane-air mixture is 810°K, at a local Mach number of the modest order 2-3, flame onset will be reached.

Combustion of gas can occur without any additional catalysts, not only because of changes in the geometry of the cross-sections of the pathways, but also because of viscous resistance of the flow.

In chapter 6 we will show that ignition of the gas can also occur because of decomposition of hydrates. The presence of hydrates on the top of offshore volcanoes is known by sampling. The presence of hydrates in onshore volcanoes can be testified to only by drilling.

All in all, it would seem that the probability of gas flaming is a parameter which varies not only from one volcano to the next but also from one eruption to the next. This probability (the ignition coefficient) can be evaluated for each volcano from the ratio of the number of eruptions with ignition to the total number of eruptions. By the law of large numbers the ignition coefficient can be evaluated most accurately for volcanoes which erupt frequently. In this study volcanoes were selected on which seven or more eruptions have been recorded; the results are given in Table 4.11. Because the ignition coefficient varies from 0 to 1, a beta-distribution was used with the data of Table 4.11 to describe the ignition coefficient statistics, giving the values $\alpha = 2.9$ and $\beta = 4$, yielding an average ignition coefficient of 42%. Figure 4.14 presents the beta-distribution and a frequency histogram of the ignition coefficient.

Table 4.11. Ignition Coefficients for Different Mud Volcanoes

Volcano	Ignition Coefficient
Lokbatan	0.421
Shikhzagirli	0.625
Keireki	0.083
Kushchi	0.454
Touragai	0.43
Makarova Bank	0.11
Bakhar	0.5
Bulla	0.33
Kumani	0.43
Nardaran-Akhtarma	0.14
Bozdag-Kobiisk	0.5

3. Flame Height Distribution

In assessing the hazard resulting from mud volcano activity during oil and gas development operations in offshore conditions, an evaluation of the flame height is of particular importance. There are about 240 records of eruptions in Azerbaijan, of which 57 have information on the flame height. According to these data the average flame height is 240 m, with a root mean-square deviation of ±50 m. The frequency histogram of Fig. 4.15 shows that low height flames predominate over tall flames, likely because the number of weak eruptions is much larger than the number of strong eruptions. Further, because there are gaps in the record with minor eruptions (and some will not flame anyway),

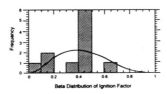

Figure 4.14. Ignition factor distribution.

therefore the estimated average flame height of 240 ± 50m must be biased by the 57 flaming eruptions reported. In chapter 5 a method for estimating parameters for mud flows will be developed. Using that method estimates of the true flame height distribution can be obtained. The approximate solution to equation (5.6) is then α=0.25 for the Azerbaijan flame height data; then θ=78m, corresponding to the distribution of all flame heights and θ=237m for the recorded flame heights, respectively.

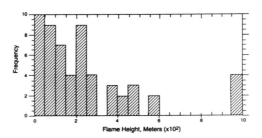

Figure 4.15. Distribution of flame heights.

4. Emitted Gas Volume Distribution

Gas emitted during eruptions bursts out in a free state, and also is given off later due to degassing of the volcanic breccia. The amount of gas given off during degassing may be evaluated from the volume of breccia; the flow of gas from gryphons and salses is measured, but the measurement of gas volumes during the active phase of an eruption has not so far proven possible. Attempts have been made to estimate gas emissions at the time of eruption (Dadashev, 1963). Mechanical calculations of the amounts of gas have been made based on data concerning height and diameter of flames, the rates of flow, and the duration of combustion. Results have been compared with experimental data obtained using a Bunsen burner. Unfortunately, the absence of data does not allow similar calculations to be made for all eruptions. Therefore results of the studies undertaken by Dadashev (1963) are used here as shown in Table 4.12.

The data were used to construct a regression curve between the rate of gas emission, R (10^6m^3/min), and the height of the flame h(m), as shown in figure 4.16,
yielding

Table 4.12. Gas Flame Data for Volcanoes

Volcano	Year	Flame height (m)	Combustion time (mins.)	Gas volume $(10^6 m^3)$	Rate of gas emission $(10^6 m^3/min)$
Bolshoi Kanizadag	1950	150	94	119	1.26
Kumani	1950	100	15	23	0.75
Touragai	1947	500	30	495	16.5
Duvannyi	1961	200	40	65	1.8
Bolshoi Maraza	1902	30	4320	120	0.03

$$R = \exp\{-0.4 + 0.006h\} \tag{4.12}$$

Using equation (4.12) for those eruptions for which there is information on the flame height and combustion time, the volumes of gases emitted during the eruptions can be estimated (Figure 4.17). Regrettably, however, it is unlikely that the flame height, other parameters of the eruptions, or the data selection, are representative. There is a high probability that small eruptions, particularly those without flames, are not accurately accounted for.

Using the beta-distribution method of the previous sub-section enables an estimate to be made of the unbiased distribution, and an equivalent scaling emission volume value of $\theta=0.59 \times 10^6 m^3$, versus the biased average value of $\theta = 36 \times 10^6 m^3$, yielding a value $\alpha=0.17$ for the maximum likelihood method, implying $(5-6) \times 10^6 m^3$ of gas emitted into the atmosphere per year during volcanic eruptions; about $10^{12}-10^{13} m^3$ during the Quaternary period.

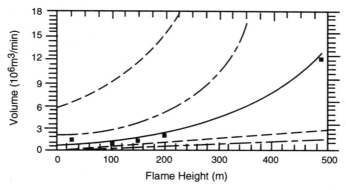

Figure 4.16. Regression of volume of emitted gas versus the height of flame.

5. Gas Volumes Emitted from Volcanic Mud

Experimental data (Dadashev, 1963) indicate that each kg of mud brought to the surface gives off about 2 cm^3 of gas. Taking the density of the mud to be 2.3 g/cm^3, the result is that each $10^3 m^3$ of rock emit 4.6 m^3 of gas. From the distribution of mud volumes measured at the surface during an eruption, it is possible to find the amount of gas given off during degassing. Because the mud volumes have an exponential distribution with an average of $120 \times 10^3 m^3$,

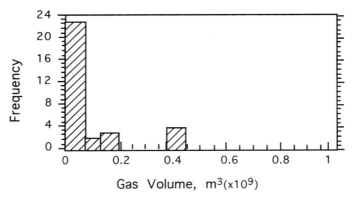

Figure 4.17. Histogram of the distribution of the volumes of free gas emitted during eruptions.

the distribution of the accompanying gas will also be exponential, with an average value of $0.6 \times 10^3 m^3$; small compared to the largest estimates ($36 \times 10^6 m^3$) from the gas flame data of Table 4.12, and a factor of a thousand less than the unbiased estimate ($0.59 \times 10^6 m^3$) of emitted gas volumes from all volcanoes.

D. Hazard Distances for Flames

Heating hazards caused by mud volcano flame eruptions are assessed here for the South Caspian Basin.

As has just been shown, the existence is well documented of flaming eruptions from mud volcanoes in both the onshore and offshore regions of Azerbaijan. Flame jets up to 1 km high are known to occur, with typical jets of 100-200 m (Lerche et al., 1996). Flame columns can be up to 50 m (and occasionally more) in diameter, can last for tens to thousands of minutes, and are known to occur not only from mud volcanoes on land or in shallow waters but also from mud volcanoes whose crests are currently several hundred meters below sea level. There is, apparently, no simple way of predicting when a particular mud volcano might flare.

In the majority of cases observed the flame column rises almost vertically but, in the presence of high winds (such as occur throughout the fall, winter and spring in Azerbaijan), the flame columns can be bent away from the vertical and, occasionally, are known to lick the surface. The potential hazard to offshore installations is clear.

The purpose of the present section is to estimate the strength of wind necessary to cause a flame column to lick the surface, and to estimate how far laterally from the flame origin such an event can occur. At the same time an estimate will be given of the temperature at the surface from flame columns that do not quite reach the surface. In this way one has available the likely scorching hazard and the likely distance away from the flame source at which scorching can occur. Because the effects are dependent on wind speed, one can also assess the most hazardous meteorological conditions for flame burning effects should flame ignition occur from a mud volcano eruption.

1. A Simple Model of Flame Column Distortion

As sketched in figure 4.18, a flame column of radius R is taken to be supplied by a steady flow of gas, of density ρ and speed V_0 at the ignition surface. The flow is taken to occur at an angle θ_0 to the horizontal so that the initial vertical component of flow is $V_0\sin\theta_0$ and the initial horizontal flow is $V_0\cos\theta_0$. Gas flow mass flux, $\dot{M}$, is then $\dot{M} = \rho V_0\pi R^2$ gs^{-1}. Take the burn of the gas to occur at some steady rate, b, so that the total length of the flame column before burnout (brennschluss) occurs is L, with a time, τ, from surface ignition to column brennschluss given by about $\tau = L/V_0$.

Suppose a downward wind of speed V_w, moving at an angle ψ to the horizontal impacts the flame column. For a viscosity ζ, the force per unit length in the lateral direction acting on the column is $F_x = \lambda 2\pi\zeta\, V_w\sin\theta\sin(\psi+\theta)$, where θ is the angle the flame column makes to the horizontal at a distance l along the column, and λ is a coefficient of order unity which allows for the non-rigid nature of the column (Taylor, 1953). The vertical force per unit column length of the wind on the column is
$F_y = -2\pi\zeta\lambda\, V_w\cos\theta\sin(\psi+\theta)$.

The gas mass per unit length, m, at a distance l along the column is $m = \pi R^2\rho$ where ρ is the gas density. Because the gas is ignited, it is buoyant relative to the surrounding air (of density ρ_a), so that the vertical motion of the mass m of gas in the column is described by

$$m\frac{d^2y}{dt^2} = +\frac{g(\rho_a-\rho)m}{\rho_a} - 2\pi\zeta\, V_w\cos\theta\sin(\psi+\theta) \qquad (4.13)$$

while the horizontal motion is given through

$$m\frac{d^2x}{dt^2} = \lambda 2\pi\zeta\, V_w\sin\theta\sin(\psi+\theta) \qquad (4.14)$$

where $\cos\theta = \dot{x}/(\dot{x}^2 + \dot{y}^2)^{1/2}$, $\sin\theta = \dot{y}/(\dot{y}^2 + \dot{x}^2)^{1/2}$ and $\dot{x}(\dot{y})$ denotes $dx/dt(dy/dt)$, respectively. With a steady-state geometry for the flame shape it is sufficient to follow the motion of the parcel of gas, of mass m, from the time (t=0) it leaves the ignition surface (x=0=y) until the brennschluss point is reached.

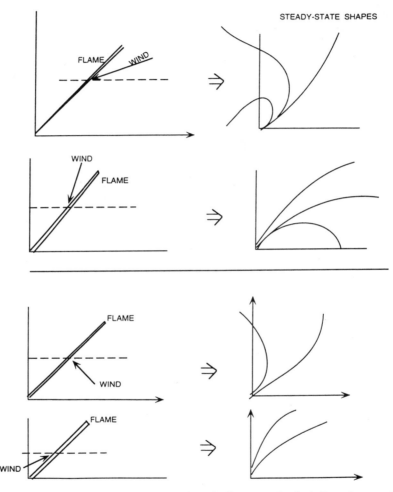

Figure 4.18. Schematic illustration of the influence of wind direction on flame column shaping.

Rewrite equations (4.13) and (4.14) in the forms

$$\ddot{y} = g\Delta - \Lambda \dot{x}(\dot{y}\cos\psi + \dot{x}\sin\psi) / (\dot{y}^2 + \dot{x}^2) \tag{4.15a}$$

$$\ddot{x} = \Lambda \, \dot{y}(\dot{y}\cos\psi + \dot{x}\sin\psi) / (\dot{y}^2 + \dot{x}^2) \qquad (4.15b)$$

where

$$\Lambda = 2\lambda\zeta V_w / (R^2\rho) \; , \; \Delta = (\rho_a - \rho)/\rho_a$$

The initial conditions on the parcel of gas are

(i) $y = 0$ on t=0;   (4.16a)

(ii) $x = 0$ on t=0;   (4.16b)

(iii) $\dot{y} = V_o\sin\theta_o$ on t=0;   (4.16c)

(iv) $\dot{x} = V_o\cos\theta_o$ on t=0;   (4.16d)

where V_o is the ejection velocity and θ_o the angle the ejection makes with the horizontal surface.

If the wind is <u>upflowing</u> rather than <u>downflowing</u> then ψ is replaced by $-\psi$ in equations (4.15a) and (4.15b); while a horizontal wind is described by $\psi=0$. All three cases are treated in Appendix A of this chapter, and lead to surprisingly different behaviors for the flame column shaping.

The general solution to equations (4.15a) and (4.15b) under the boundary conditions (4.16a)-(4.16d) is not difficult, but is tedious, and so has been relegated to Appendix A. However, certain factors can be deduced immediately. Suppose that the flame has reached its highest point ($\dot{y}=0$) and is about to turn over to descend towards the originating surface. Then, from equation (4.15a), at any place where $\dot{y}=0$ one has

$$\ddot{y} = g\Delta - \Lambda\sin\psi \qquad (4.17)$$

But if $\ddot{y}$ is positive then $\dot{y}$ is increasing and one cannot have a turn-over. Hence one requires $\ddot{y} < 0$ when $\dot{y}=0$. Thus: in order to have a turn-over at all one requires a downward vertical wind speed ($V_w\sin\psi$) greater than a critical value minimum, V_*, given by

$$V_w\sin\psi \geq \frac{g(\rho_a-\rho)R^2}{2\lambda\zeta} (\rho/\rho_a) \equiv V_* \qquad (4.18)$$

Thin (R small) and low buoyancy (ρ close to ρ_a) flame columns can be more easily "bent" by a low speed wind than can large, highly buoyant, flame columns which require a higher critical downward wind speed, as expected.

A rough estimate of the critical wind speed, V_*, can be made as follows. Let the wind impacting the flame column be heated to the flame temperature (at least 900°K for methane burning). Then, relative to cold air (of density $\rho_a \approx 10^{-3}\text{gcm}^{-3}$ at a temperature of 300°K), the heated air density is

$$\rho/\rho_a \cong 300/900 \sim 1/3 \tag{4.19}$$

The effective viscosity, ζ, is not the molecular viscosity, but rather the turbulent viscosity of heated air disturbed by the presence of the flowing flame column. Simple arguments (Parker, 1978) suggest that the effective viscosity, ζ, is then given by about

$$\zeta \approx \frac{1}{10} v_{Th} \cdot R \, \rho \tag{4.20}$$

where R is the flame column radius and v_{Th} is the thermal speed of gas molecules ($\approx 1.2 \times 10^5$ cm/sec). It then follows that a rough estimate is

$$V_* \cong \frac{g\rho_a(1-\rho/\rho_a) \, (\rho/\rho_a).10R}{2\lambda \, v_{Th}\rho} \quad \text{cm/sec}$$

$$\approx 5gR(1-\rho/\rho_a) / (\lambda \, v_{Th}) \quad \text{cm/sec}$$

$$\approx R/(50\lambda) \quad \text{cm/sec.} \tag{4.21}$$

For a 10 m radius flame column, one has $V_* \cong (0.8/\lambda)$ km/hr. Thus downward wind speeds in excess of around 1-5 km/hr can have a significant impact on a 10 m wide flame; and, for a 50 m radius flame the corresponding critical windspeed is around 10-50 km/hr. A worst case hazard scenario would then suggest that any sustained wind of a few km/hr should be treated as a danger point should flame ignition of a mud volcano occur.

2. Heating Hazards

The dynamics of a wind-modified flame column provide the background for assessing the hazard from flame heating of points on the surface-presumed to be where the exploration and production platforms are located. In Appendix B, the technical details are given for assessing the values of temperature induced on the surface by a distorted flame column as a function of position along the surface from the emergence point of the flame. Here we concentrate on a few typical examples to illustrate the magnitude of variation of "hot spots". For simplicity we take it that any surface temperature above 60°C (333°K), which is stinging hot against human flesh, produces a thermal hazard.

Estimates of flame temperatures are relatively easily made from several directions. First, the minimum self-ignition temperature for a methane-air mix is about 530°C (~800°K) and is higher as the air humidity is increased.

Second, field observations, made by both authors near "perpetual flames"[1] in onshore Azerbaijan, did not permit the authors to stand stationary closer than about 50 meters to the flames (of order 1 meter high), and the flame color ranged from yellow to bright blue; both factors indicate a flame temperature of around 800-1000°C. Third, historical records and photographic evidence, show partial melting of low-grade steel used in some on-shore and off-shore rigs that were exposed to flames, suggesting temperatures of around 800-1500°C. Fourth, for the largest flames (of order 1-1.2 km high), it is recorded that it was bright enough at night in Baku to read a newspaper by the bluish light of the flame, again indicating flame temperatures of around 800-1500°C. We will therefore, carry through the case histories using a flame column temperature of 1000°C as representative of conditions. Because the radiative flux is geometry dependent, it is then easy to scale results to any other flame temperature. Thus: for a surface temperature $T_*(x_0)$ produced by a flame of temperature 1000°C (see Appendix B), then a flame of temperature T_f will produce a surface temperature $T_1(x_0)$ given by

$$T_1(x_0) = (T_*(x_0)/1273) \, T_f$$

where $T_*(x_0)$, T_f and $T_1(x_0)$ are all in °K, and x_0 is the lateral distance in the plane of the wind and flame measured from the surface emergence point of the flame; and all other flame and wind conditions are kept the same in both cases. Hence, one merely scales the results of any one calculation to obtain other situations at different flame temperatures.

The situation is not, however, so easily scaled for other parameters due to the highly non-linear nature of the dependence of the equations on flame radius, wind speed and direction, flame emergence direction, and emergence speed; as well as on burn-time (brennschluss) for a steady-state flame, which determines flame length. However, interest centers on estimating likely worst case hazards so that rough ranges of some parameters can be provided. For instance, a methane flame at 1000°C means a thermal particle speed of about 1400 m/s; the dynamical flame speed at surface emergence is unlikely to exceed this value. And, further, observations indicate the establishment of kilometer-high flames in around a few seconds, indicating emergent flame dynamical velocities of order a few hundred m/s. Accordingly, we have chosen to perform most of the case histories at a flame emergence speed of 300 m/s.

To provide a worst case hazard estimate we have also run the case histories with a constant flame radius of 50 m, which is typical for very large flames, but which is likely an overestimate for smaller length (100 m, say) flames. However, it is better to overestimate the worst case hazard than underestimate it and, anyway, it is presumably the largest flames which are the most hazardous, so one errs on the side of caution this way.

---

[1] "Perpetual flames" are those for which the earliest written records indicate the flames have existed for at least several hundred years without quenching once.

The brennschluss time is, arguably, one of the main controlling parameters on flame dynamics - if the brennschluss time is very short then the flame burns out over a short distance, while a long brennschluss time enables the effects of the wind on flame-shaping to be maximally portrayed. Observations of flames indicate lengths of several hundred meters to just over a kilometer, so the brennschluss time must be of order a few seconds. We have run case histories with different brennschluss times, but in the second to just over 10 second range.

For the prevailing wind there are two parameters: the strength of the wind and its direction relative to the surficial emergence angle of the flame. A downward (upward) component to the wind will, in general, attempt to deflect a rising flame column downward (upward), except when the angle of the wind to the horizontal is smaller than the emergence angle of the flame when an upthrust is applied to the flame, as depicted in the schematic illustration of figure 4.18, causing a reverse "bending" to the flame, with the amount of the reversal dependent on the wind strength.

Case histories include different wind angles for both upward and downward conditions of the wind, and also different wind speeds, so that one can obtain an appreciation of the dynamical ranges involved. First, a set of results is presented representative of conditions appropriate to the recorded flame eruption of 1961 on Duvanny Island, where eye-witness observations were made close to the flame. In this way one has some idea of the way model behaviors and observations conform. Second, three groups of case histories are presented, organized with respect to initial emergence angle of the flame, in order to provide some idea of the spatial scale of worst-case hazards.

3. Test Cases

a. The Eruption of Duvanny Island in 1961

The surviving eye-witnesses reported an almost vertical flame eruption of 200-300 m height, lasting about 45 mins., and, prior to the flame, only a light breeze blowing (as also indicated from the newspaper meteorological records of the day). The rapid (few minutes?) time frame over which the barracks (at about 1.5 km distance from the flame) had paint blistered and mattresses smoldering argues, as discussed previously, for a "hot" flame, so we use 1000°C.

The major unrecorded values for the flame are its radius and its methane/air ratio. By increasing the brennschluss time we can allow for a longer flame given the same emergence flame speed, and so a larger radiative flux, thereby increasing surface temperature. By increasing the flame radius, we can have a higher surface temperature at a larger distance from the flame site. For the "light breeze" conditions reported, we have taken a horizontal wind speed of 5 km/hr as not atypical.

Shown in figure 4.19a-4.19d are typical results for the model. Figure 4.19a records the height and bending of a 20 m radius flame which has a brennschluss time of 0.65 sec., corresponding to a flame length of 194 m, for a flame erupting at 80° to the horizontal; and also recorded is the surficial temperature (in °C) as a function of lateral position measured from the flame emergence point. Note that at 1.5 km from the flame, the surficial temperature is about 150°C, and the "safety distance" to drop below 60°C is about 2-4 km to either side of the flame. If one takes an ignition temperature for smoldering mattresses to be the same as the temperature at which book-paper burns (about 451°F ($\approx$ 231.7°C) according to Bradbury (1957)), then either the brennschluss time or the flame radius need to be slightly increased relative to the values used in producing figure 4.19a.

Figure 4.19b records the corresponding results when the brennschluss time is increased to 1.1 sec with all other parameters the same as for figure 4.19a. Now note that the flame length is about 327 m, the safety distance is increased to around 3 km, and, at 1.5 km from the flame, the surficial temperature is increased to 200°C, still slightly less than the 231.7°C required. A point of diminishing returns is reached now because, while increasing the brennschluss time will provide greater surficial temperatures, it will also increase the flame column length beyond 327 m - which would violate the eye-witness observations of around 200-300 m high flame.

Thus, in figure 4.19c the radius of the flame is increased to 50 m, with a brennschluss time of 0.72 sec, thereby providing a flame length of 216 m. The safety distances are now 3.8 km on the upwind side of the flame and 4 km on the downwind side; while the surficial temperature at 1.5 km from the flame is approximately 250°C, precisely in the correct region.

Indeed, increasing the brennschluss time to 1.13 sec. produces a flame column of 340 m, a safety distance of 4.8 km, and a temperature of about 300°C at 1.5. km - slightly too high.

Thus it would appear that the Duvanny Island flame eruption is encompassed by a flame radius of greater than about 20 m and only a little bit less than around 50 m, and a brennschluss time of around 0.7 sec. Other parameter values will not produce the required temperature of around 232°C at the observation point 1.5 km from the source, as well as also yielding a flame length in the 200-300 m range as estimated by the surviving eye-witnesses.

And then the "stinging hot against human flesh" temperature of 60°C is achieved at a distance beyond about 4.8 km from the flame, suggesting that around a 5 km radius from the flame is a safety distance.

b. Case History A: High Angle Flames

While the Duvanny Island flame eruption is, perhaps, representative of the typical hazard conditions caused by flame heating (because most flames are tens to a few hundred meters high), of greater concern is to estimate a likely worst case hazard. While the probability of such a worst case is small, the fact that such hazards can (and do) occur provides a basis for determining the likely conditions one should be prepared to encounter.

To this end, the historical record provides observations of flame columns reaching to a height of about 1 to 1.5 km, and with radii variously estimated to reach around 50 m, and lasting from several minutes to more than

98

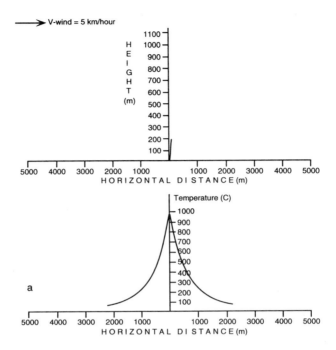

Figure 4.19. Influence of a light (5km/hr) wind on a flame column considered representative of the Duvanny Island 1961 event: (a) a flame column of 200 m length, and 20 m radius, together with the corresponding surficial temperature;

4,000 minutes (Bagirov et al., 1996). So, in discussing the three groups of case histories under variable wind conditions, attention is generally focused (but not exclusively) on flames of 50 m radius, and with brennschluss times of several seconds, corresponding to typical flame lengths of around 1-1.5 km (the flame height may be much smaller than the flame length if considerable bending of the flame occurs). In addition the flame temperature is taken to be at 1000°C throughout all examples.

For flames which erupt close to perpendicular to the surface, the effect of a wind with an upward component larger than the horizontal component is, by-and-large, to increase the vertical tendency of the flame. The consequences are to make the temperature at the surface more akin to the case of zero wind. In this first group of high angle case histories we shall, therefore, concentrate on winds which are horizontal or which have downwardly directed components because such winds produce significant flame distortions.

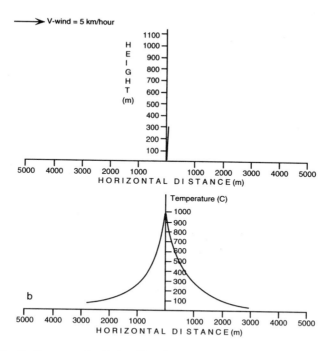

Figure 4.19. Influence of a light (5km/hr) wind on a flame column considered representative of the Duvanny Island 1961 event: (b) as for figure 4.19(a) but with the column length increased to 330 m.

(i) Flames at 70° to the Horizontal

Figures 4.20a-4.20c show the flame distortion caused as a wind blowing at 40° to the horizontal with a downwardly directed component increases in strength from 50 km/hr (Figure 4.20a), through 70 km/hr (Figure 4.20b) to 100 km/hr (Figure 4.20c), with the total length of the flame in the range 1.5 ± 0.3 km. A 50 km/hr wind has only a slight tendency to bend the flame (of radius 30 m) as shown in figure 4.20a, and the surficial temperature pattern is almost symmetric with a safety distance to reach less than 60°C of 8.2 km to the left of the flame and 7.8 km to the right.

 As the wind speed is increased to 70 km/hr there is now a bend-over of the flame produced (figure 4.20b) with a slight distortion of the surficial temperature, and safety distances of 8.8 km (left) and 8 km (right). Increasing the wind speed even more to 100 km/hr forces a considerably bending of the flame, as depicted in figure 4.20c, to the point that the flame tip is now only 200 m above the surface. Correspondingly, there is a major change in the surface temperature because more of the flame length is closer to the surface than in the case of a low speed wind (figure 4.20a). To the left of the emergence point there is a secondary maximum in temperature due to the close approach of the flame tip to the surface and, overall, there is a significant

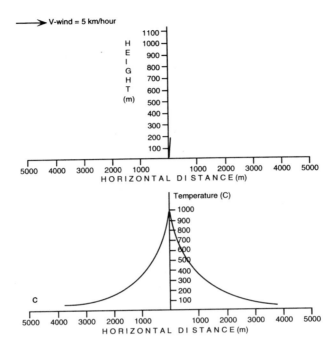

Figure 4.19. Influence of a light (5km/hr) wind on a flame column considered representative of the Duvanny Island 1961 event: (c) as for figure 4.19(a) but with a 50 m radius, and a 200 m column length.

increase in temperature to both the left and right of the emergence point, with safety distances now of 11.4 km (left) and 9.6 km (right). Thus, _increases_ of 3.2 km (left) and 1.8 km (right) in the safety distances are produced by the 100 km/hr wind relative to the 50 km/hr wind - which already called for substantial safety distances of around 8 km from the flame emergence point.

(ii) Flames at 80° to the Horizontal

The patterns shown in figures 4.20a-c for flames at 70° to the horizontal are virtually identical for flames at all higher angles, and so are not shown here. Instead we concentrate now on a different aspect of the hazard problem. Suppose the flame radius is now either 20 m or 50 m rather than the 30 m of the figure 4.20 cases. Then the viscous drag of the wind is changed because the turbulent viscosity is proportional to the radius of the flame but the cross-section of the flame is also changed, so that the relative importance of the viscous drag diminishes for a larger radius flame. Thus, such larger radius flames are harder to "bend" with a given wind speed, all other factors being equal.

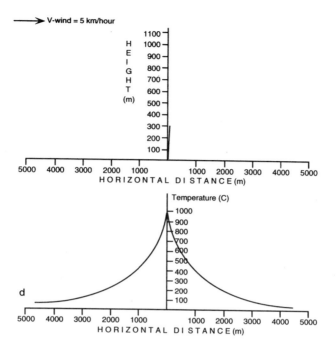

Figure 4.19. Influence of a light (5km/hr) wind on a flame column considered representative of the Duvanny Island 1961 event: (d) as for figure 4.19(c) but with the column length increased to 340 m.

For instance, figure 4.21a shows the bending of a 20 m radius flame in a 100 km/hr wind, which can be compared directly with figure 4.21c for a 30 m radius flame. The "bending" occurs at a smaller height in figure 4.21a and drives the flame closer to the eruption surface than for the 30 m flame of figure 4.21c. Also of concern is the original ejection speed of the flame column. Figure 4.21b shows a thin (30 m radius) flame column ejected at 1080 km/hr, while for comparison figure 4.21b shows the greater bending of a thicker (50 m radius) flame column in the same wind but when the ejection speed of the flame is lowered to 720 km/hr. The higher bending of the thicker flame column in figure 4.21b implies a higher surface temperature and, indeed, the safety distance to the left (right) of the emergence point of the flame is 10.8 km (9.8 km) for the flame of figure 4.21b versus 8 km (8 km) in the case of the thinner flame of figure 4.21b.

Thus: it is not only the flame radius, and wind speed and direction which can play significant roles in raising the surficial temperature, but the emergence speed of the flame also provides a sensitive control on flame heating.

These two groups of examples indicate the various scale patterns of heating effects for close to vertical flames under high hazard conditions.

102

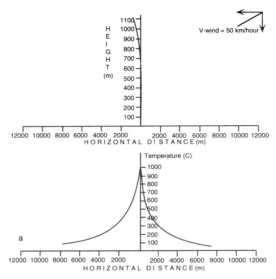

Figure 4.20. Flame distortion and surficial temperature heating caused by a wind blowing at 40° to the horizontal with a downwardly directed component increasing in strength from: (a) 50 km/hr. The flame emergence angle is 70° to the horizontal, the radius is 30 m; and flame length is 1.5 ± 0.3 km.

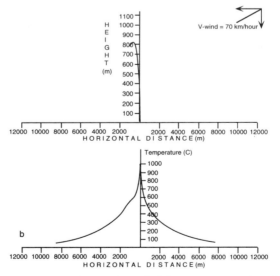

Figure 4.20. Flame distortion and surficial temperature heating caused by a wind blowing at 40° to the horizontal with a downwardly directed component increasing in strength from: (b) 70 km/hr. The flame emergence angle is 70° to the horizontal, the radius is 30 m; and flame length is 1.5 ± 0.3 km.

Roughly: about 8-10 km lateral distance from such flames represents a safety distance for any downward directed wind speed in excess of about 50 km/hr.

c. Case History B: Low Angle Flames

When the emergence angle to the horizon of the flame is low, then the direction of the wind can play a significant role in either raising or lowering the surficial

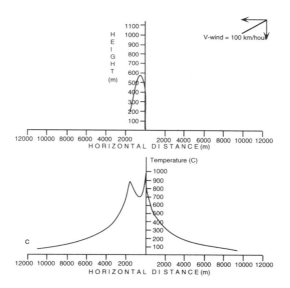

Figure 4.20. Flame distortion and surficial temperature heating caused by a wind blowing at 40° to the horizontal with a downwardly directed component increasing in strength from: (c) 100 km/hr. The flame emergence angle is 70° to the horizontal, the radius is 30 m; and flame length is 1.5 ± 0.3 km.

temperature caused by flame heating. To illustrate this point case histories were run in which a 50 m radius flame emerges at 30° to the horizontal, but in which the direction of the wind (of fixed speed 50 km/hr) varies from being downgoing at 50°, 30°, and 10° to the horizontal, respectively, and upgoing at 1°, 20° and 70° to the horizontal, respectively. Very different temperature effects are produced by the two groups of cases.

104

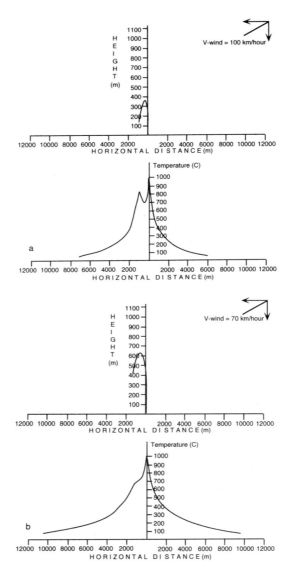

Figure 4.21. Flame distortion and surficial temperature heating for flames emergent at 80° to the horizontal in a 100 km/hr. wind, partially downwardly directed. (a) a 20 m radius flame; (b) a 50 m radius flame.

Figures 4.22a,b, and c show the results for a downgoing wind. For a wind at 50°, figure 4.22a indicates that the flame is completely bent over and "licks" the emergent surface at about 1.8 km from the emergence point. The corresponding surficial temperature profile of figure 4.22a indicates that the safety distance is 9.8 km on the left of the flame, and 11.4 km to the right, as measured from the emergence point. For a more horizontal wind at 30°, figure 4.22b indicates that the flame now is bent and only "licks" the emergence surface again at 2.4 km from the emergence point, creating a larger region of higher surficial temperature as indicated by safety distances of 11.6 km and 14 km to the left and right of the flame emergence point, respectively.

As the wind becomes even more horizontal at 10°, figure 4.22c indicates that the flame reaches brennschluss before it can be completely bent over. But the greater height of the flame, and greater length of flame column, now heats the surface even more, with safety distances of 13.2 km (left) and 16.4 km (right).

By way of comparison, a wind of identical strength (50 km/hr) but with an upgoing component will tend to deflect a low angle flame more towards the vertical and so diminish the heating effect at the surface. Figures 4.23a,b, and c illustrate this effect. For an almost horizontal wind (1° upward) but directed into the flame, figure 4.23a shows the flame column being vertically deflected prior to brennschluss being reached at a flame column length of 1.3 km. The corresponding surface temperature in figure 4.23a shows safety distances now of 9.2 km (left) and 10 km (right). As the upward angle of the wind is increased to 20°, figure 4.23b shows a flame column which is even more vertically directed than for figure 4.23a, with safety distances now reduced to 9.0 km (left) and 9.9 km (right). By the time the upward wind angle is 70°, the flame column is turned almost vertically, with safety distances of around 9 km (left) and 9.8 km (right).

Thus, the influence of upgoing or downgoing equal strength winds on low angle flames is extremely different. The downgoing wind components provide a greater surficial temperature hazard out to larger distances from the flame emergence point.

E. Conclusions

Mud volcanism is a unique natural phenomenon, with an activity pattern of alternating between the slow emission of relatively small amounts of solid, liquid and gaseous products from deep crustal layers, and short-lived eruptions, during which massive amounts of breccia and high volumes of gas (which sometimes ignite) are released. The formation of new mud volcanoes is a rare event. Over the past 100 years only four new mud volcanoes have been observed onshore in Azerbaijan. Furthermore, newly-formed mud volcanoes are usually very small, and their eruptions do not present a serious hazard. The most probable waiting time for the formation of a new volcano in the Chirag area is about 1600 years. The probability that a new volcano will not form in this area over the next 100 years is greater than 0.9.

106

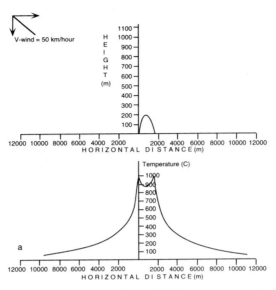

Figure 4.22a. Flame distortion and surficial temperature heating for flames emergent at 30° to the horizontal and of 50 m radius. The wind speed is fixed at 50 km/hr but the wind direction is variable. Downgoing wind results at 50°.

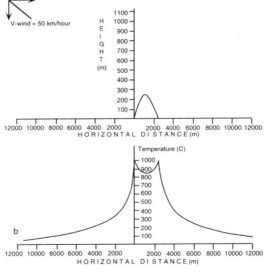

Figure 4.22b. Flame distortion and surficial temperature heating for flames emergent at 30° to the horizontal and of 50 m radius. The wind speed is fixed at 50 km/hr but the wind direction is variable. Downgoing wind results at 30°.

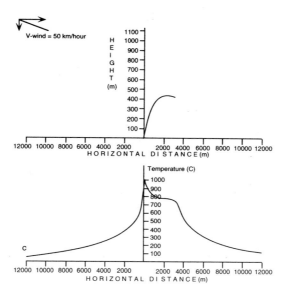

Figure 4.22c. Flame distortion and surficial temperature heating for flames emergent at 30° to the horizontal and of 50 m radius. The wind speed is fixed at 50 km/hr but the wind direction is variable. Downgoing wind results at 10° to the horizontal are depicted.

In the Chirag area, where two mud volcanoes exist, the formation of new gryphons and salses is probable. These gryphons usually lie within the area of the crater field and along faults. In the volcanoes studied the average diameter of the crater field is about 120 m. The eruption frequency of mud volcanoes can be described by an exponential distribution. The most probable waiting time for a first eruption is about 95 years, and for a powerful or average-strength eruption is about 270 years.

Large volumes of gas are emitted during eruptions. The gas volume distribution can be described by an exponential distribution with an average value of about $590 \times 10^6 m^3$. Small volumes of gas are also given off by degassing of volcanic breccia, but the mean value is only about 560 m^3, many orders of magnitude lower than the gas volume emitted during the eruptive phase. The gas composition is mostly methane (averaging 95%), with smaller amounts of carbon dioxide (3.5%), and minor quantities of nitrogen, ethane, heavy hydrocarbons, hydrogen and helium. The hydrogen sulfide content is extremely low, of order of 10^{-5}%.

Gas ignition depends both on volume and on the rate at which gas moves along exit channels and is emitted from such channels. The ignition coefficient, characterizing the probability that the gas will self-ignite during an

108

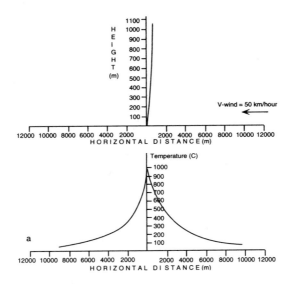

Figure 4.23a. As for figure 4.22 but with the wind direction changed to upward at 1°.

eruption, is described by a beta-distribution with parameters $\alpha=2.9$ and $\beta = 4$, and about a 42% average likelihood that gas will flame spontaneously. The flame height, which can reach hundreds of meters, averages 77 m, and there is a 95% probability that a flame will be in the range 60 m to 100 m, although in the majority of eruptions there is no flame recorded. To estimate the heating hazard distance the dynamical modeling of section D is needed.

Apart from the specific examples presented here, which were chosen to illustrate the major points as simply as possible, we have run a large variety of other cases with different parameters for the flame column (radius, brennschluss time, emergence angle, flame temperature, initial emergence velocity) and for the wind (direction with or against the flame, speed, upgoing/downgoing component angle).

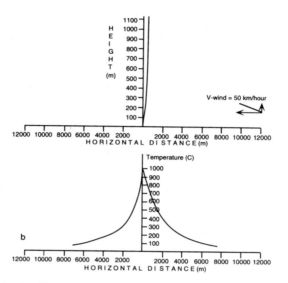

Figure 4.23b. As for figure 4.22 but with the wind direction changed to upward at 20°.

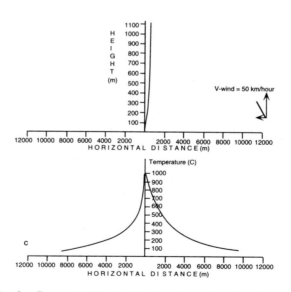

Figure 4.23c. As for figure 4.22 but with the wind direction changed to upward at 70° to the horizontal.

In general, both the high angle and low angle flame columns cause stinging hot heating at the emergent surface out to of order 10 km from the flame emergence point, for wind speeds in excess of about 50 km/hr, with the low angle flames providing a greater hazard closer to the flame than the high angle flames.

It would seem that, irrespective of conditions, safety distances for flames of several hundred meters length are at least of order 5 km and, more often, of order 10 km.

While there is, as yet, no way known to predict when a particular mud diapir will flame, nor any apparent relation between the resulting flame radius, flame total lifetime, breccia ejecta, and the size, shape, or location of a particular diapir, nevertheless it is clear from the historical record that the occurrence of flaming eruptions can be a major hazard in the onshore and offshore regions of Azerbaijan. One of the concerns is not just the appearance of such flame columns (reaching heights recorded of up to 1.2 km) but their hazard for rigs, infrastructures (e.g. pipelines), and personnel safety. The influence of winds, which blow prevalently in the South Caspian region throughout about nine months in the year, is, therefore, a factor to be considered in influencing flame shapes and the corresponding surficial heating.

As might have been anticipated, either roughly horizontal winds, or winds with a downward component, provide an increased hazard for any flame columns relative to winds with upward components.

For large flame columns, dangerous conditions occur for wind speeds in excess of about 40-50 km/hr, when it is possible to "bend" flame columns so they come close to "licking" the surface. The corresponding temperature hazards set lateral safety distances of around 8-15 km from the emergence point of the flame to reach less than stinging hot (60°C) conditions against human flesh, with the smaller safety distances of around 8 km corresponding to flame columns of a few hundred meters, and 15 km to columns of around 1 km.

It would seem that this hazard potential is one that should be of some concern to exploration and production operations in both the onshore and offshore regions of Azerbaijan.

Appendix A - Solutions to Equations (4.15a) and (4.15b) under the initial conditions of equations (4.16)

A. Downward Flowing Wind ($\sin\psi > 0$)

Write $\dot{y}=u$, $\dot{x}=v$, when equations (4.15a) and (4.15b) can be written, respectively, as

$$\Lambda^{-1}\dot{u} = (V_*/V_w) - v\,(u\cos\psi + v\sin\psi) / (u^2 + v^2) \qquad (A1)$$

$$\Lambda^{-1}\dot{v} = u\,(u\cos\psi + v\sin\psi) / (u^2 + v^2) \qquad (A2)$$

with $\qquad$ $u = V_o\sin\theta_o$ on t =0, $\quad v = V_o\cos\theta_o$ on t=0. $\qquad$ (A3)

Divide equation (A1) by equation (A2) to obtain

$$du/dv = \frac{r(u^2+v^2) - v(u\cos\psi + v\sin\psi)}{u(u\cos\psi + v\sin\psi)}$$ (A4)

where $r = V_*/V_w$.

To solve equation (A4) write

$$u = vA(v)$$ (A5)

when A(v) then satisfies the equation

$$v\frac{dA}{dv} + A = \frac{r(A^2+1) - (A\cos\psi + \sin\psi)}{A(A\cos\psi + \sin\psi)}$$ (A6)

Equation (A6) can be re-arranged to read

$$dA.A(A\cos\psi+\sin\psi) (A^2+1)^{-1}(A\cos\psi+\sin\psi-r)^{-1} = -dv\ v^{-1}$$ (A7)

which has the solution

$$v = V_o\cos\theta_o \ [\cos^2\theta_o(A^2+1)]^\alpha \left[\frac{\tan\theta_o\cos\psi+\sin\psi-r}{A\cos\psi + \sin\psi-r}\right]^\beta x$$

$$\exp\ \{r\cos\psi\ (r-\sin\psi)(1-2r\sin\psi + r^2)^{-1}\ (\tan^{-1}A - \theta_o)\}$$ (A8)

where

$$\alpha = - (1-r\sin\psi) / (2(1-2r\sin\psi+r^2))$$ (A9a)

$$\beta = - r(\sin\psi-r) / (1-2r\sin\psi+r^2)$$ (A9b)

Hence, one can also express the vertical component of velocity, u, through

$$u = Av(A)$$ (A9)

The position of the mass element at any instant of time is then given parametrically, with elapsed time, t, through

$$t(A) = \Lambda^{-1} \int_A^{\tan\theta_o} dz\ v(z)\ (z\cos\psi+\sin\psi-r)^{-1}$$ (A10)

and coordinate positions of the mass through

$$y(A) = \Lambda^{-1} \int_A^{\tan\theta_o} dz\ zv(z)^2\ (z\cos\psi+\sin\psi-r)^{-1}$$ (A11)

and

$$x(A) = A^{-1} \int_A^{\tan\theta_o} dz\, v(z)^2 (z\cos\psi + \sin\psi - r)^{-1} \qquad (A12)$$

As A decreases towards zero from the initial value of $\tan\theta_o$, equation (A9) indicates that the vertical velocity, u, heads to zero, representing the largest height reached by the flame.

If $\sin\psi > r$, i.e. $V_w\sin\psi > V_*$, then the factor $A\cos\psi + \sin\psi - r$ is positive on $A=0+$, and stays positive until A reaches the negative value $A_o = -(\sin\psi - r)/\cos\psi$, at which point the factor approaches zero from above. But, since the factor $A\cos\psi + \sin\psi - r$ occurs as a denominator in equations (A10), (A11) and (A12), it follows that if A ever reaches A_o from above then $t\to\infty$, $y\to\infty$, and $x\to\infty$. Thus there is a negative value of A, closer to zero than A_o, for which y eventually goes to zero - the flame has reached the horizontal surface. If $\sin\psi = r$ (the critical separation case), then $A\cos\psi + \sin\psi - r$ is just zero on $A=0+$; thus there must be a positive value of A (smaller than $\tan\theta_o$) at which the brennschluss point is reached because, for this case, $y(A)$ cannot reverse sign, so the flame is never bent past the horizontal. If $\sin\psi < r$, then the factor $A\cos\psi + \sin\psi - r$ goes to zero at the positive value $A_1 = (r-\sin\psi)/\cos\psi$, at which value there is still a positive upward velocity for the flame. If $\sin\psi \le r \le \sin(\psi+\theta_o)/\cos\theta_o$, then $A_1 \le \tan\theta_o$, while if $r > \sin(\psi+\theta_o)/\cos\theta_o$, then $A_1 > \tan\theta_o$. Thus, A decreases (increases) away from the initial value of $\tan\theta_o$ for $r > \sin\psi$, depending on $r\sin(\psi+\theta_o)/\cos\theta_o (>\sin(\psi+\theta_o)/\cos\theta_o)$.

B. *Horizontal Flowing Wind ($\sin\psi=0$)*

In this case equation (A8) reduces to

$$v(A) = V_o\cos\theta_o[\cos^2\theta_o(A^2+1)]^{-\alpha_o}\left(\frac{\tan\theta_o-r}{A-r}\right)^{\beta_o} x$$

$$\exp\left[r^2(1+r^2)^{-1}(\tan^{-1}A-\theta_o)\right] \qquad (A13)$$

with

$$\alpha_o = -\frac{1}{2}(1+r^2)^{-1} ; \beta_o = -r^2/(1+r^2) \qquad (A14)$$

while the corresponding parametric representatives for time, t, and coordinates reduce to

$$t(A) = A^{-1} \int_A^{\tan\theta_o} dz\, v(z) / (z-r) \qquad (A15a)$$

$$y(A) = \Lambda^{-1} \int_A^{\tan\theta_o} dz \, zv(z)^2 / (z-r) \tag{A15b}$$

$$x(A) = \Lambda^{-1} \int_A^{\tan\theta_o} dz \, v(z)^2 / (z-r) \tag{A15c}$$

It is not possible to have A decrease below r in this case because then $v(A)$ would go to zero on $A=r$, as would the vertical speed, but time would go negative. Hence, either A tends to r from above, or A increases from its initial value of $\tan\theta_o$ (which keeps $v(A)$ positive as required by equation (4.15b)). Thus the vertical velocity is always positive in this case and there is no bend-over of the flame - although it can be "stretched" a long way in the horizontal direction.

C. *Upward Flowing Wind* $(\sin\psi<0)$

In this case write $\Psi = -\psi$, with $\Psi > 0$, when equation (A8) becomes

$$v(A) = V_o\cos\theta_o \, [\cos^2\theta_o(A^2+1)]^a \left[\frac{(A\cos\Psi - \sin\Psi - r)}{(\tan\theta_o\cos\Psi - \sin\Psi - r)}\right]^{-b} x$$

$$\exp\left[r\cos\Psi(r+\sin\Psi)(1+2r\sin\Psi+r^2)^{-1}(\tan^{-1}A - \theta_o)\right] \tag{A16}$$

where

$$a = -(1 + r\sin\Psi) / [2(1+2r\sin\Psi+r^2)] \tag{A17a}$$

$$b = r(\sin\Psi+r) / (1+2r\sin\Psi+r^2) \tag{A17b}$$

and where time and position coordinates are given parametrically through

$$t(A) = \Lambda^{-1} \int_A^{\tan\theta_o} dz \, v(z)^2 (z\cos\Psi - \sin\Psi - r)^{-1} \tag{A18a}$$

$$y(A) = \Lambda^{-1} \int_A^{\tan\theta_o} zd \, zv(z)^2 (z\cos\Psi - \sin\Psi - r)^{-1} \tag{A18b}$$

$$x(A) = \Lambda^{-1} \int_A^{\tan\theta_o} dz \, v(z)^2 (z\cos\Psi - \sin\Psi - r)^{-1} \tag{A18c}$$

In this case note that A must always exceed $(\sin\Psi+r)/\cos\Psi$ in order to keep $v(A)$ real, so that $\tan\theta_o \geq \tan\Psi + r\sec\Psi$ is required in order to have any flame column. The column must be close to the vertical direction in this case and, as A increases, so too does $y(A)$. Thus there is no bend-over possible of the flame column, just a slight bend in the horizontal flow direction, but the column is always vertically rising, and it is required to start on $y=0$ (the horizontal surface for emission) at a high enough angle to the horizontal to maintain $\tan\theta_o \geq \tan\psi + r\sec\Psi$ in the steady-state case.

D. The Length of the Flame Column

If the burning of gas in the column occurs at a constant rate then the time to burn all the gas is fixed at say $t = \tau$. The length, L, of the flame part of the column is then derivable from

$$\int_0^{x_{max}} \frac{(1+(dy/dx)^2)^{1/2}dx}{(\dot{y}^2 + \dot{x}^2)^{1/2}} = \tau \tag{A19}$$

where x_{max} is the lateral distance prior to brennschluss. Equation (A19) can be rewritten[2]

$$\int_0^{x_{max}} \frac{dx}{\dot{x}} = \tau \quad \equiv \quad \int_0^{x_{max}} \frac{dx}{v(A)} \tag{A20}$$

so that the flame column length, L, is

$$L = \int_0^{x_{max}} (\dot{x}^2 + \dot{y}^2)^{1/2} \frac{dx}{\dot{x}} \quad \equiv \quad \int_0^{x_{max}} (A^2+1)^{1/2}\, dx \tag{A21}$$

Hence, once x_{max} is determined from equation (A20) it is a simple matter of quadratures to determine the flame column length, L, from equation (A21).

Appendix B - Heating Effects

Because the flame column length, and the shape of the flame, have been calculated in Appendix A, the heating by the flame of a particular point on the horizon $y=0$ can also be estimated. Two components of heating should be considered: radiation and convection. The distance, D of a unit length of flame

---

[2]Note that $dx=\Lambda^{-1}\, dA\, v(A)^2(A\cos\Psi+\sin\psi-r)^{-1}$ in the downflowing wind case, so the integrals can be rewritten in terms of the parametric variable A.

column, currently occupying coordinates $(x(A), y(A))$ from a particular point x_o (in the plane of the flame column) on the horizon $y=0$, is given by

$$D(A) = [y(A)^2 + (x_o - x(A))^2]^{1/2}$$

(B1)

A. Radiation

The luminosity, L, of a volume element is

$$L = \sigma T^4 \qquad (\text{erg cm}^{-2}\text{sec}^{-1})$$

(B2)

where σ is the Stefan-Boltzmann constant and T is the flame temperature. The surface area of the element, of length dl, is $2\pi Rdl$ so that the output of radiation is

$$\text{Rad} = \sigma T^4 \, 2\pi Rdl$$

(B3)

At a distance D ($>$R) from the element the received flux of radiation, F, is less than about

$$F = \sigma T^4 \, 2\pi Rdl/4\pi D^2 \qquad \text{erg cm}^{-2}\text{sec}^{-1}$$

(B4)

From all elements along the flame column (taking them all to be at about the same temperature), the total flux of radiation impinging on the point x_o is

$$F_{total} = \frac{1}{2} R\sigma T^4 \int_0^L \frac{dl}{D^2} \qquad , D \geq R.$$

(B5)

which integral can be written in terms of the lateral position, x, along the flame column as

$$F_{total} = \frac{1}{2} R\sigma T^4 \int_0^{x_{max}} \frac{(A^2+1)^{1/2}dx}{[y(A)^2 + (x_o-x(A))^2]} \qquad \text{ergcm}^{-2}\text{sec}^{-1}, \text{ in } D \geq R.$$

(B6)

If that total flux is picked up as a heating event of the surface then, eventually, the surface would rise to a temperature T_* such that the flux of radiation from the surface (σT_*^4) exactly balanced the flux being received from the flame column, F_{total}. The temperature T_* is then given by

$$T_*(x_0) = T\left\{\frac{1}{2}R \int_0^{X_{max}} \frac{(A^2+1)^{1/2}\,dx}{[y(A)^2 + (x_0-x(A))^2]}\right\}^{1/4} \quad in\ D \geq R.$$

(B7)

$$= TG$$

Of course, because the surface temperature, T_*, must be less than or equal to the flame temperature (second law of thermodynamics), the limiting factor on this approximate estimation of temperature is the value of G, which must be less than or, at most, equal to unity.

The temperature $T_*(x_0)$ calculated solely from the radiative flux is a minimum estimate, because heated air driven by the wind will also contribute to total heating at x_0 as follows.

B. Convection

As well as the flux of direct radiation reaching the point x_0, there is also a flux due to the wind carrying energy. For air of density ρ_a, traveling with the wind at speed V_w, and with the air having been heated to temperature T by passage across the flame front, the convective flux of energy is

$$F_{conv.} = \rho_a TCV_w \qquad\qquad erg\,cm^{-2}sec^{-1} \qquad\qquad (B8)$$

where C is the specific heat of air.

From an element of heated air of radius R and length dl at a position $(x(A), y(A))$ the total convected flux is

$$F_{total} = \rho_a TC2\pi Rdl.\ V_w \qquad\qquad erg\,sec^{-1} \qquad\qquad (B9)$$

The relative contributions of convective versus radiative transport are difficult to evaluate without a detailed flow model of heated air, surficial interaction of heated air with the surface, etc. For this reason in the body of the chapter we have restricted the estimate of surficial heating by the flame column to the contribution from radiative flux - not only is this the easiest component to calculate but it also provides a minimum estimate of surficial heating. If the surface temperature, T_*, produced by radiative heating is enough on its own to provide a danger factor to surficial equipment, then the convective factor will just make the situation that much worse. Thus a minimum hazard assessment is produced.

Appendix C - Summaries of Some Historical Flame Observations

From Sultanov, A.D., Agabekov, M.G., "Makarov Bank mud volcano eruption",
Doklady Akademii Nauk Azerbaijanskoi SSR, vol. XV, 1959, No. 2 (in Russian)
pp. 143-146.

During the 20th century eruptions on the Makarov Bank were recorded in 1906, 1912, 1917, 1921, 1925, 1933 and 1941. For the previous century the eruption of 1876 is known. The eruptions were associated with the ejection of gas and mud. The cones which formed during eruption sometimes rose a couple of meters higher than the ocean surface and formed islands, which then disappeared as the sea currents washed out these masses of mud until the base of current flow was reached, leading to a submarine bank. This happened in 1876, 1921 and during the last eruption of 15 October 1958, which is the subject of the paper. The sea depth here is significant and, as the authors remarked, if this volcano would be onshore, then a cone a couple of hundred meters high would form. (The authors, however, did not notice that such hills onshore were formed as a result of hundreds or even thousands of eruptions, while on the Makarov Bank and other offshore eruptions, islands of a hundreds meters length scale appear after a single eruption. The mud, being sloppy, should flow even faster in water than hill growth on land. Therefore, island appearance and disappearance is most likely due to the growth and buoyancy of the top of the diapir. This diapir rises because of expanding gases and then, after the eruption, deflates again. The ejected mud plays a secondary role in island formation).

The eruption of 1958 was different from the previous eruptions because of the strong flame occurrence. The flame had a height of 200-250 meters and was observed from Baku and Oil Rocks, which were 80 km away. (From other sources the flame was observed from Akhsu pass (about 200 km away)). The captain of the ship Fataliyev watched the fire from the beginning. About 10 p.m. he was at a distance of 27 km from the eruption center. But it seemed that the fire was very close and they changed the direction of the ship to the south. A huge surface area of the water was on fire. Some flashes of flame moved mostly in the direction of the wind, which blew that day from south-west to north-east. (This is probably an error because in the next page the authors indicate the opposite direction of the wind. And the ship therefore did not have trouble with the flame heat. They were on the upwind side). Then the ship continued its way to Baku Bay. When they were about 18 km from the eruption center the crew felt that the deck was covered by sand. But it was night. And 30-35 minutes after beginning the flame was extinguished. The next day the crew found all the deck covered by black sand. The density of the cover was approximately 1 handful from 1 m². The grain sizes were from decimals of millimeter to 2 millimeters. The grains crunched under foot pressure because they were hollow, like cored spheres. That probably happened, when the sand particles were sucked into the flame. Under high temperature they blew out due to expansion of inner moisture, and sometimes burst. There are small bubbles on the surface of the sandstone particles. Such particles probably are erupted on other volcanoes as well. But, because they are very small, people did not notice them.

The interesting thing was that the wind blew from north-east to south-west (most likely) and the vessel was to the north-east from Makarov Bank at that time (that fact was mentioned a second time), and was at a distance of 18 km from the eruption center. The authors explain the observed phenomena by the fact that wind can have different directions at different heights. While the air strata near the surface had a NE-SW direction, the strata above (at the heights 1000-1200 m) can have an oppositely directed flow. So the small particles could be lifted by the flame column to higher levels and then transported over far distances. The next day a group from the Academy of Sciences visited this place. The water over a significant area was turbid, which indicated an intensive wash-out process. The group did not see the island, but in their opinion the eruption could happen only from an island, because only in such a case could the sand particles be scattered in the air.

Yakubov, A.A., Dadashev, F.G., Magerramova, F.S. "Eruption of Ayrantekyan Mud Volcano", Doklady Akademii Nauk Azerbaijana, vol. XXI, 1965, N2, pp. 33-37.

In the night from 6 to 7th of October 1964 Ayrantekyan volcano erupted. An eruption was observed from the drilling rig, approximately 1 km away from the eruption center. A muffled rumble was followed by explosion and a flame of height 100-150 m arose over the volcano. The area around the volcano was illuminated to a distance of 10-15 km. The flame was wider in the upper part and had a shape of turned-over cone. The fire burned 20-30 minutes and then got weaker. Its height decreased to 20-100 m and the shape had more regular contours. The temperature in the rig increased and it was hard to breathe. At the beginning the flame was whitish and later black smoke appeared. When the mud started being thrown, the intensity of the fire decreased; 4-5 hours after the eruption its height was 10-20 m and then later 5-2 meters. An eruption center crater had a diameter 250-450 m. To October 8, the height of flame remained 1.5-2 m, but even on October 21, the gas still burned (0.5 m height).

The ground to the north of the crater (500-600 m away) was burned and had dark-brown color, while the surface of the fresh mud was reddish-brown. Bushes were burned from the flame side of the volcano. According to the melted breccia, and comparing with lab analysis, the temperature of burning gas was 1200-1400°C. The burning gas seeped from different fractures as well. In the central part of the crater there were some holes in the erupted mud, formed by falling rocks. The shapes of the holes are different from round to almost rectangle. Diameters of the holes were 30-200 cm and depths 20-100 cm.

Different mud flows formed. The largest one had a length of 1600-1800 m and width 30-50 m. Another had a length 400-450 m and width 200-250 m. Total area of the mud cover was 175,000 m^2 and of thickness 2-6 m. A lot of fractures and faults occurred. The displacement of the largest fault was 2m with the depth of the open part 0.5-0.7 m. Intensive fracturing was also near the old crater field, where the gryphons and domes are. Fractures here destroyed some domes of 1 m height. The whole area was covered by rock fragments from 1 cm to 1 m in size. The largest fragment had a size 0.5 x 0.8 x 1.3 m³. A few fragment can be found outside the mud cover.

Gorin, V.A., "Eruption of Bolshoi Kanizadag Mud Volcano", Doklady Akademii Nauk Azerbaijana, vol. VI, 1950, No. 7.

May 12, 1950 an eruption of Bolshoi Kanizadag occurred. During the prior 150 years there was no activity of this volcano. I watched this eruption from the very beginning and then compared my observation with those of other geologists, who also watched the eruption.

The eruption started about 6 a.m. An explosion first and then a flame appeared at 6:15 a.m. However, most likely, breccia ejection happened before. Because when I reached the edge of the mud flow at 8:15 a.m. there was already a dry crust on it and the velocity of the mud flow was very slow, which shows that the mud ejection happened some time at night.

The first stage of the fire lasted from 6:15 a.m. to 7:51 a.m. There was enough time to draw the picture (The picture in Gorin's paper showed that the height (not the length) of the flame is approximately one third the height of the 400-450 m high volcano). There was a strong force 4-5 north wind and the angle of the flame to the horizon reached 45° and the height 150-200 meters. The picture was drawn from a point 4-5 km to the south-east from the volcano. The flame column consisted of the shooting up every second of reddish-white puffs of fire, which rotated like fire balls. The smoke had an ash-grey color. At 7:51 a.m. the fire and smoke stopped. The rumble stopped as well. Such a temporary silence was maintained during all the time that I climbed up the hill.

At 8:30 a.m. I stayed at the edge of the mud covered area, about 200 meters from the crater. Every minute pieces of mud were thrown to a height of 10-15 meters. Going around I saw spots of completely burned grass and wormwood, surrounded by areas covered with grass and not affected by heat action. Such burned areas were mostly on the south and west sides from the crater. It seemed that the rapidly flying fire balls just touch the ground in some places. From the leeward side of the crater it smelled of sulfur and burned grass.

At 8:47 a.m. gas started to erupt again. It came out with low level noise and ignited at a height of approximately 5 meters and then formed those fire puff-balls. Those balls came periodically, like pulsations. At the same time the pieces of thick mud were thrown to a height of 5-10 meters. Such conditions kept going for 12 minutes until 8:59 a.m. This phase was quiet and allowed one to watch an eruption from a distance of 300 meters. The burning gas puffs reached diameters of 10 meters at a height of 15-20 meters and then got smaller and disappeared. From 8:59 to 9:31 a.m. the gas eruption stopped. In 32 minutes the gas emission and ignition started again. But 6 minutes later a strong pushing out of the mud column occurred. With a high rumble fragments of rock were thrown to a height of 70-80 meters during 15 minutes. Burning gas was not observed that time.

At 9:52 a.m. suddenly a silence came. It seemed that the crater was blocked up by the rocks, but from 10 a.m. mud ejection started again and stopped only at 12:30.

As one can judge, the eruption was characterized mostly by gas emission, rather than mud ejection. Mud cover does not exceed 1.5 m thickness and it did not cover even the whole top of the volcano.

A number of fractures formed (shown on the picture). To the south from those fractures the old mud cover had fallen by 3 meters.

Yakubov, A.A., Gadjiyev, Y.A., Matanov, F.A., Atakishiyev, I.S., "Eruption of Kelany Mud Volcano", Doklady Akademii Nauk Azerbaijana, vol. XXVI, 1970, No. 5, pp. 55-60.

December 12, 1969 the people of the village Polatly were witnesses of the eruption of Kelany volcano, taking place at 8:40 p.m. It started with a high rumble and tremors. Three to five minutes later a strong explosion happened and the flame of 350-400 meters height illuminated the area for 40-50 km around. About 10 minutes later a huge mass of mud erupted. The flame had an orange-red color with puffs of black smoke. Some rock fragments were thrown to a height of 150-200 meters and fell down not only on the erupted mud flow, but also far from that place (1 km away). Twenty minutes later the intensity of the volcano reduced. The height of the flame dropped by a factor two. But the volcano continued to throw rocks to a height of 45-60 meters and burning gas came from six different vents.

At 9:15 p.m. the rumble and the tremors got stronger, followed by a deafening explosion, and the flame column rose to a height of 250-300 m. The mass of erupted mud increased. Such explosions repeated at 10 p.m., 10:15, 11:10, 11:25 and 11:35 p.m. Very intensive burning lasted 5-15 minutes, then the intensity dropped. The last gas eruption (11:35) was the strongest. The flame reached 400 meters height.

By December 15 the flame height in the center of the crater was 8-10 m, and then reduced, and for a couple of days was 0.3-0.5 meters. The erupted mud covered 70,000 m^2 with an average thickness between 3-8 m. As a result of the fallen stones holes on the mud surface formed with diameters 3-5 meters and depths 1-2 m. In the center of the volcano burned places (10 m x 25 m) were found. There are a number of such places.

In the north-east part there was a fractured area of width 120-130 meters. Vertical displacement reached 2.5 meters and one of the fractures extended for 2-3 km.

Zhemerev, V.S., "Eruption of Lokbatan mud volcano", ANKh, 1954, N11, pp. 30-31.

July 30, 1951 (22:22 p.m.) that was a strong eruption on Lokbatan. A crater was observed 2 days before an eruption, and no other evidence was noticed. We were at the foot of volcano 1-2 minutes before the eruption. First an underground rumble similar to far peals of thunder and hissing of gas were heard. Almost at the same time we heard a strong explosion. That was caused by the inflammation of gas. The flame column had a height of 400-500 meters and diameter 40-50 meters. The crater surface of 50 m x 100 m was on fire. The flame tongues reached 10-15 meters. The flame was visible in Baku, Alyat (about 70-80 km away). An area of 3-6 km in radius around the volcano was lit up as in daytime. Twelve minutes after the beginning of eruption the flame dropped very fast and then gas burned on an area of 30 m x 70 m and formed separate tongues of 10-15 meters height. We approached the crater in 15 minutes. The grass around the crater at a distance of 100-120 meters continued burning. It was difficult to breathe, probably because of high concentration of CO_2 and CO. The temperature of erupted mud was then approximately 40-50°C.

The rock fragments of 2 to 5 kg mass and a lot of smaller fragments were distributed around a radius of 100 meters. The rocks falling to the ground cracked and then scattered to a distance of 1-1.5 meters, which indicates that they fell down from a considerable height.

The first minutes after the eruption a cloud of dark-grey color arose over the flame to a height of 1.5-2 km. It was formed by steam, smoke and small clay particles. Those particles returned and formed a "clay rain", which started 20-30 minutes after the eruption started and lasted 1 hour and 30 minutes. At a distance of 3 km the volcanic dust condensed on the roads and crunched under the feet, which was noticed the next day. This "dust" consisted of baked clay particles of different shapes and 2 mm and less in size. The color was mostly light-grey and dark-brown.

All night and the next day gas burned, but the flame was weaker. The next day July 31 at 4:30 p.m. the volcano erupted again. This time an eruption occurred from the side crater. The flame had a height of 60-70 meters and burned for 2 minutes. Puffs of black smoke were over the flame. Particles of different sizes were pushed to a height equal to a third of the flame height. No mud was ejected this time. During the eruption fractures of 1 m width formed in an area of 400 meter width. The subsidence of the blocks in some places reached 2 meters. The subsidence was noticed on the morning of July 31 and this process finished August 3. The surface burning continued until August 11 and then it continued underground and in fractures. As a result the mud burned and had a brick color. At the beginning of the eruption blue-white smoke came from the fractures and smelled of chlorine and sulfur.

Sultanov, A.D., Dadashev, F.G. "Eruption of the Mud Volcano on Duvanny Island", Izvestiya Akademii Nauk Azerbaijana. Seriya Geology, Geography and Oil, 1962, N3, pp. 73-81.

On September 4, 1961 a strong eruption occurred on Duvanny Island at 8:45 a.m. Muffled underground rumble. Ground slowly lifted 10-15 meters. Then black smoke appeared, which was followed by a flame of 200-250 meters height. The temperature around increased and wooden constructions at a distance of 300 meters away inflamed. On the upper part a mushroom shape cloud formed, which started to condense in 5-10 minutes and the temperature decreased. The flame however burned for 30-45 minutes.

The diameter of the erupted mud cover was 400-600 m and the thickness was 5-7 meters. The mud was almost dry. Spherical empty rocks ("lapilli") were found on the coast about 30 km away from the island.

Yakubov, A.A., Salayev, S.G., "Eruption of Bozdag Kobiiski Mud Volcano", ANKh, 1953, N12, pp. 3-4.

An eruption occurred on August 23, 1953. The previous eruption was 60 years before. At 10:30 a.m. a strong underground rumble started. As a result of the explosion a crater field of 400 meters in diameter subsided to 1 meter. The fractures formed with most of the fractures having a concentric character and surrounding the crater. The width of the largest fracture was 1 meter, displacement 1.5 meters and depth 3-4 meters. At distances of 7 and 25 meters from a particular fracture 2 other concentric fractures formed (width 20-30 cm). Some linear fractures (up to 40 cm width) were in the crater.

In the southeast part of the crater gas was emitted. The flame height was 1 meter and lasted until the first rain. During 7 days the crater subsided. August 24 the displacement on the main fracture was 0.8 m; on August 31 it was 1.5 m. The width of that fracture increased during that time from 20-25 cm to 55-60 cm.

Nadirov, S.G., Zeinalov, M.M., "A New Eruption of Bozdag (Kobiiski) Mud Volcano", ANKh, 1958, N3, pp. 9-10.

On August 27, 1957 at approximately 10 p.m. Bozdag volcano erupted again. A small flame burned 25-30 minutes. A thick mud flow had a length of 150 m and width of 5-6 m. The thickness was up to 1.5-2 meters in the central part. Some very small fractures formed.

Buniat-Zade, Z.A., Gorin, V.A., "About one eruption of gas-oil Volcano Alyat-Cape (March 20, 1967)", Doklady Akademii Nauk Azerbaijana, vol. XXIV, 1968, N9, pp. 29-34.

(This is now called Bahar volcano).

First started with tremor, then rumble and at 9 p.m. a huge orange-red flame of 100-200 meters height appeared. About 2 million m^3 mud was taken to the surface. Two days later there were two small flames of 2-2.5 meters height. Ten meters around the grass was burned. Empty spherical burned rocks of large pea-size were around.

5 Mud Flow Hazards

Mud volcanoes are known to erupt and spread massive amounts of breccia over a scale of tens of kilometers, both as airborne ejecta and as mud flows. During eruptions over several hours or days, massive amounts of breccia and large volumes of gas are ejected. Breccia flows, several meters thick, spread over hundreds of meters (occasionally several kilometers) in length. Flowing mud can destroy all platforms, pipelines and other operational equipment, because mud volcanoes are usually associated with oil and gas fields.

Therefore, it is important to know the potential lengths and possible directions of mud flows for each volcano to plan safe locations of future platforms. The length of a mud flow, as well as its thickness and direction, depend on: 1) volume of mud erupted; 2) the morphology of a mud volcano and the position of each eruptive center on the volcano; 3) the transport medium in which the eruption occurs (offshore versus onshore; air versus water).

The kinematic and dynamic characteristics of flows differ depending on the medium in which the flows erupt and move. For onshore volcanoes, where the flow erupts onto the earth's surface in air, mud <u>flows</u> possess high viscosity and density, and have a low rate of movement. The form of the flow, and its velocity, then depend on the topography (angle of slope) of the surface and, particularly, on the moisture content of the erupting mass which controls the viscosity. The flow speed is also influenced by rainfall and season of the year. The results of measurements of flow rates (Potapov, 1935) show that mud flow velocities increase after rain. The duration of the flow movement may vary from several days to several months, and depends on how quickly the mud dries. The period of flow movement until complete stoppage is affected by climatic factors. Air at high temperatures, and also wind, lead to the drying of mud flows and a slowing of their rates of movement. Low air temperatures (frost) also slow the rate of flow movement due to mud freezing, with subsequent restoration of the movement after a period of warming.

Mud flows on land differ from those in the marine environment primarily because the boundaries of the media (mud and air) are clearly defined on land but are blurred in marine conditions (mud with sea water). Onshore the leading edge of the mud flow becomes more viscous in the process of moving forward and drying out, and acts as a braking mechanism (Fig. 5.1a), resulting in the mud flow being unbroken and continuous.

In marine conditions the mud flow from the eruptive center is more dense and viscous than lower down the slope or in the near-surface parts of the

123

124

flow. A fraction of the mud, together with bubbles of emitted gas, creates a penumbra of turbid mud/water, which moves downslope with the mud flow (Fig. 5.1b).

The volume of mud erupted by any volcano in a given basin can be described by a random process variable with parameters which can be estimated by statistical methods. The same methods allow one to describe a distribution of size parameters of mud flow.

Despite a thorough search of the literature and catalogued data on mud-volcano eruptions in Azerbaijan, only two references have been found concerning the length of submarine mud flows; but copious information has been found on the dimensions of mud flows on land for undertaking statistical

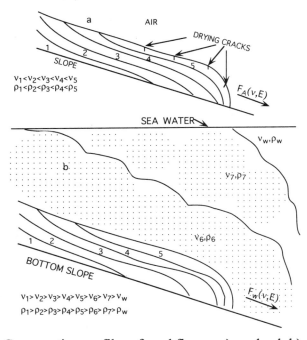

Figure 5.1. Comparative profiles of mud flows: a) on land; b) offshore.

analysis. Accordingly, the parameters investigated for their probability distributions are: length, width and thickness of flow, area of the mud cover, and the volume of erupted mud.

Analysis of the linear characteristics of mud flows observed on land has shown that they are mostly associated with surface topography. If an eruptive center lies in a flat area, then the mud volcano flows are isometric in form and cover approximately circular areas. If there is a topographic slope, the mud flows tend to be directed towards the maximum slope angle and acquire an elongate form. The same may also be true with submarine flows, because the mud flow in an aqueous medium acquires an elongate linear shape, even with relatively gentle angles of slope, due to its lower viscosity (Garde and Ranga Raju, 1978).

It would seem that the most important factors affecting the length of marine flows are:

(1) In offshore conditions a part of the mud volume will be expended in forming turbid flows, thereby decreasing the remaining volume of mud flow and therefore its probable length as shown on Fig. 5.1b;

(2) On the other hand, in view of the lower viscosity of mud flows in marine conditions than on land, the probable distance covered by a mud flow on the same angle of slope will be greater.

It is not possible, on the basis of the limited data available, to determine which of these two factors will most influence the probable length of mud flows. To estimate the scales of the mud flow in submarine conditions, a 3-D mud flow model, called MOSED3D, has been used to examine the mud flow problem. The method is based on the concept that a mud current moves along a three-dimensional surface according to the balance between gravity (driving force) and friction (resistance force) in the fluid media, supported against topographic facies resistance.

Modeling of mud flow and deposition is accomplished by taking quanta of mud, released at a suspected eruptive center, and allowing the mass of mud breccia to flow, constrained by the existing topography of the mud volcano slope and by previously deposited mud flows of earlier eruptions. Each quantum of released mud is transported downslope and deposited when its flow energy drops below a critical value. The mud flow can also cause erosion of the basal sediments and of previous mud flow deposits on the slopes of the volcano; the total mass of the mud current then follows the transport rules. Each quantum of mud can be composed of variable fractions of different lithologic types, ranging from very fine-grained to coarse-grained material. Deposition at a given location takes place according to the distribution of fractions that reached that location: coarse-grained material is deposited first and fine-grained material last.

The Chirag oil field area in the Caspian Sea has been chosen to demonstrate application of the method. A map of the sea bottom in the area surrounding a mud volcano was digitized, and mud flows from possible eruptive centers (and with variable volumes) modelled, indicating amounts, directions, and thicknesses of potential flows - of concern in rig siting in the offshore South Caspian waters.

A. Distribution of Variables related to Mud Volcano Flows

Predicting the lengths of mud volcano flows is important because, during an eruption, moving mud has destructive strength. In marine conditions the destructive path may endanger offshore rig installations with foundations in the path of the mud flow. Therefore, it is important to study the probable distributions of lengths, areas, and directions of mud-volcano flows, in order that lower risk to installations can be provided, perhaps by siting rigs at safer distances from an eruptive center.

Figure 5.2 sets out provisional <u>linear</u> parameters of mud flows. The length of a flow (L) is taken as the distance between the most widely-spaced points on the same flow. The width of the flow (W) is taken as the distance between opposite sides of the flow, perpendicular to the length. The radius along the length of the flow (R_L) is the distance between the eruptive center to the farthest point on the flow. The radius across the width (R_W) is the distance from the eruptive center to the edge of the flow, perpendicular to the length of the flow. Ideally, these parameters are related to one another by:

$$R_W = W/2 \qquad (5.1a)$$
$$R_L - L - R_W = L-W/2 \qquad (5.1b)$$

Using the prior statistical analysis of length L, width W, area A, thickness H, volume V, the radius along the length, R_L, and the width, R_W, can easily be derived. Histograms of the distributions of numbers of observed flows versus the different calculated parameters are given in Figs. 5.3, 5.4, 5.5 and 5.6, from which it is seen that the average value of the radius along the length is about 200 meters, and about 70 meters along the width. It is also about 95% certain that R_L does not exceed about 225 m, and that R_W does not exceed about 85 m. These results are provisional because much of the data characterizing small flows is absent.

To estimate the distribution of the true values of these parameters, two suppositions have been made: (i) Let the distribution, p(h), of a parameter value, h, be exponentially distributed as

$$p(h) = (1/\theta) \exp\{-h/\theta\} \qquad (5.2)$$

where θ is a fixed scale value. This assumption honors the requirement that the probability of a small flow is much greater than that of a large flow; (ii) Let the probability of a gap in information on flow size increase with a decrease in the flow dimensions. For n eruptions, introduce a vector indicator of the gap, R = $(R_1, R_2, ..., R_n)$, each of the elements having the value 0 or 1 depending on whether there is a gap in the data on flow size or an observation, respectively. The conditional distribution of recording a flow of size y_i is

$$P\{R_i = 0|y_i, \theta\} = (y_i)^{\alpha-1} \qquad (5.3a)$$

$$P\{R_i = 1|y_i, \theta\} = 1 - (y_i)^{\alpha-1} \qquad , \text{ in } 0<\alpha<1, \; y_i >1 \qquad (5.3b)$$

in terms of a fixed parameter, α, characterizing the statistics of flow size. Now the maximum likelihood for the parameter θ in equation (5.2) is given by:

$$\theta = \sum_{i=1}^{n} y_i /(m+(n-m)\,(1-\alpha)) \qquad (5.4)$$

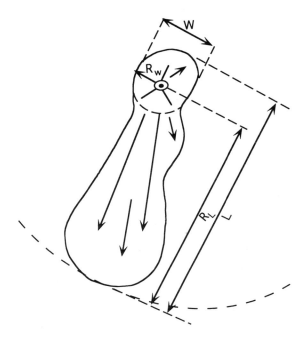

Figure 5.2. Linear parameters of a mud volcanic flow.

where n is the number of known eruptions; m is the true number of eruptions, and the recorded size of the ith flow is y_i. The actual value of the parameter, θ, must then lie within the range:

$$\sum_{i=1}^{n} y_i /n < \theta < \sum_{i=1}^{n} y_i /m \qquad (5.5)$$

The maximum likelihood value of α is then obtained from:

$$\sum_{i=1}^{n} (-y_i^{\alpha-1}) \log(y_i)/(1-y_i^{\alpha-1}) + (n-m) \log (\sum_{i=1}^{n} y_i)-\log(m+(n-m)(\alpha-1))+ (n-m)\psi(\alpha)=0$$

$$(5.6)$$

where $\psi(\alpha)$ is the psi function, (Gradshteyn and Ryzhik, 1965).

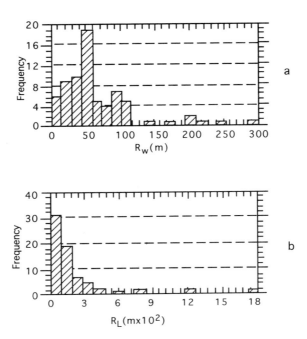

Figure 5.3. Histogram of the distribution of parameters of mud flows: a) R_W; b) R_L.

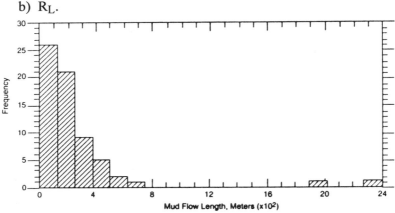

Figure 5.4. Histogram of the distribution of the values of length of mud flow.

The approximate solution to equation (5.6) for R_L and R_W is $\theta_L \approx 75$m and $\theta_w = 30$m respectively, and for volumes of breccia is $\theta_v = 120 \times 10^3 \text{m}^3$ (the sample mean of this value is $315 \times 10^3 \text{m}^3$). Taking into account the estimated number of eruptions in Azerbaijan as 10 per year, the assessment of the volume of breccia brought to the surface from deep zones by mud volcanoes is about

1.2x10^{6}m^3, implying that during the Quaternary period the volume of mud flowing from volcanoes is of order 10^{12}m^3, which is of the same order as all marine Quaternary sediments in onshore Azerbaijan.

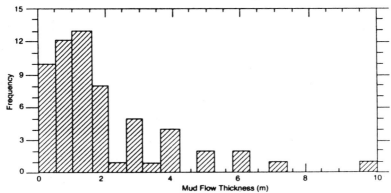

Figure 5.5. Histogram of the distribution of the values of thickness of mud flow.

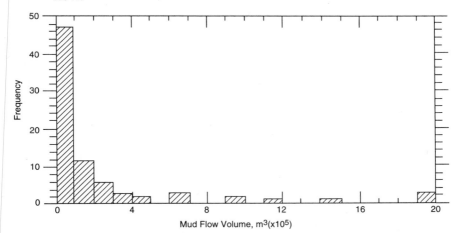

Figure 5.6. Histogram of the distribution of the values of volume of mud flow.

B. Dependence of Variables related to Mud Volcano Eruptions

In contrast to the parameters of onshore eruptions, submarine eruptions are much more difficult to measure. It is sometimes possible to measure only a single parameter. Given individual parameters, it is then important to be able to predict other parameters. Correlation and regression analyses have been done to establish quantitative relations between eruption parameters of ejecta, such as the length (L), width (W), thickness (H), and area (S).

A classical calculation of covariance and correlation matrices would seem the appropriate procedure. However, note that the matrix of mud volcano observations contains omissions in the data, preventing a classical solution.

Accordingly, for each pair of parameters where data are available for <u>both</u> parameters, the covariance and correlation coefficients have been calculated.

Correlation analysis shows that the most closely related parameters are those describing mud-volcano flows. The most important statistical relationships between (L,W,H,S and V) are:

$$L = 1.2V + 126 \tag{5.7a}$$

$$W = 0.64V + 91 \tag{5.7b}$$

with correlation coefficients of 0.9 and 0.92, respectively, with data taken from 49 and 48 observations, respectively. Other characteristics of the regression model (standard error, sum of squares etc.), and other graphs of regression curves, are presented in Figs. 5.7 through 5.13.

In calculating the mud-volcano flow volume, V, the most important role is played by the thickness, H, of the flow with

$$H = 10^{-0.64}V^{0.28} \tag{5.8}$$

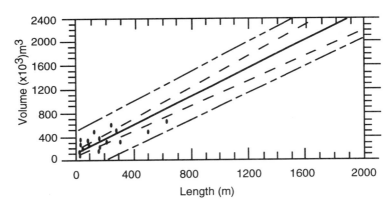

Figure 5.7. Regression curve of the length of mud flow versus volume.

with a correlation coefficient of 0.71 (Fig. 5.8). Equation (5.8) is of use for planning field development when mud volcano breccia have been penetrated in the well section, because predictions of the ejecta volume and, therefore, area of distribution, together with length and width of the ancient mud volcano flow can then be made.

A linear model provides a less satisfactory correlation of the form:

$$H = 1.52 + 0.0028V \tag{5.9}$$

with a lower correlation coefficient of only 0.59 (Fig. 5.10).

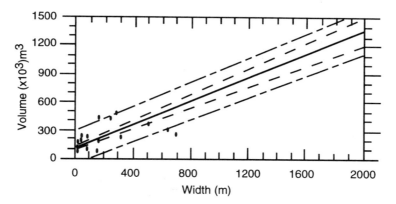

Figure 5.8. Regression curve of the width of mud flow versus volume.

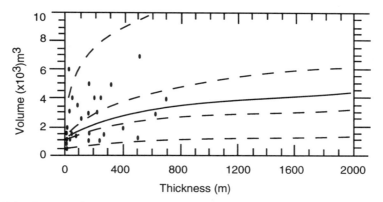

Figure 5.9. Regression curve of the thickness of mud flow versus volume for the power law model.

Other relationships between length and width of flow, length and area of cover, and also width and area are (Figs. 5.11, 5.12, and 5.13):

$$L = 1.5W + 23 \tag{5.10a}$$

$$L = 10^{3.6}A^{0.6} \tag{5.10b}$$

$$W = 1.4A + 87 \tag{5.10c}$$

Equations (5.7 through 5.10) have practical value for predicting unknown parameters of mud flows from known parameters.

132

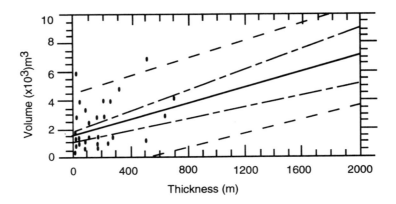

Figure 5.10. Regression curve of the thickness of mud flow versus volume for the linear model.

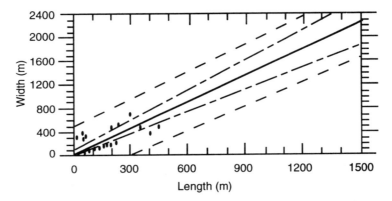

Figure 5.11. Regression curve of the length of mud flow versus width.

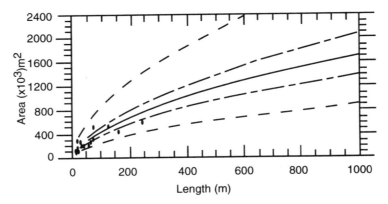

Figure 5.12. Regression curve of the length of mud flow versus area.

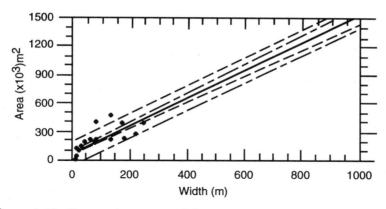

Figure 5.13. Regression curve of the width of mud flow versus area.

C. Prediction of Possible Scales of Mud Flows Based on Dynamical Models

The estimation of mud flow parameters from the historical data, shown above, has been done for all volcanoes of Azerbaijan. This assessment can be used as a first approximation, especially for onshore volcanoes. As has been noted, it is not yet clear how good those assessments are for offshore volcanoes.

However, analysis of the linear characteristics of mud flows observed on land has shown that they are mostly associated with surface topography. If an eruptive center lies in a flat area, then the mud volcano flows are isometric in form and cover approximately circular areas. If there is a topographic slope, the mud flows tend to be directed towards the maximum slope angle and acquire an elongate form. The same may also be true with submarine flows, because mud flow in an aqueous medium acquires an elongate linear shape, even with relatively gentle angles of slope, due to its lower viscosity (Garde and Ranga Raju, 1978). Therefore the form and morphology of a volcano is very important for the estimation of mud flow.

Morphologically onshore mud volcanoes occur as hills and raised areas, sometimes reaching upwards of 400 m in height, and with volumes of up to several x 10^7 m^3. Externally, mud volcanoes are similar to magmatic volcanoes, often displaying a dome-like structure. The products of mud-volcano activity are carried to the surface along exit channels, leading to craters at the surface. The crater field is most commonly circular to oval in outline and is surrounded by one or more concentric crater ramparts. The crater forms an area of subsidence, and varies in form from gently convex to a deep caldera. The area of the crater plateau can reach 10 km^2, and the crater rampart may rise 5-25 m above the center of the crater. The morphology of mud volcano breccia flows depends mainly on the topography and on the breccia composition.

Volcanoes lying offshore differ morphologically from those on land, forming islands (7 such islands currently occur in the Caspian Sea) or submarine banks. At the time of an eruption some volcanoes form new islands, which are often eroded within several days (see Chapter 10).

134

The aim of this section is to show the track of a mud flow of given volume, erupted on a volcano with known morphological structure. The quantitative method and associated program (MOSED3D) described in Cao and Lerche (1994) have been used to address this problem. The method assumes a "slump" deposition of mud on the surface, with the release of "quanta" of mud being triggered by internal catastrophic failure (much as a snow avalanche). The consequent development of turbidite sequences is but one instance where episodic and catastrophic "pulsing" of sediment models is required. The computer model MOSED3D simulates the flow of sediment in three dimensions as a result of a sudden release of a quantum of sediment.

The basic outline of the problem to be modeled is sketched in figure 5.14. A quantum of sediment or mud is released over a given volume centered at a location with higher potential energy on a basin slope. The ejected mud mixes with fluid (water, air, etc.), forming a gravity-driven current flowing downslope. The mud will then be transported and deposited; slope sediments can also be eroded when the mud current is strong enough.

In order to simulate the transport, deposition and erosion of mud and sediment during such a mud current, the following assumptions and conditions are used:

    # Failure of the slope (mud ejection) is autochthonous (no external energy is provided) and instantaneous;

    # Mud current is constrained by the existing topography of the slope;

    # Basin is filled with fluid (water, air, etc.), which is initially completely still (no eddy currents);

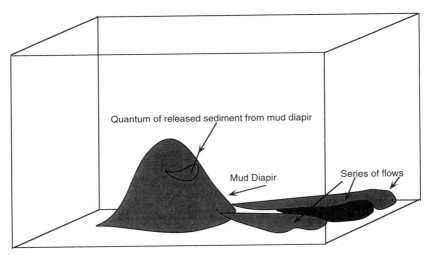

Figure 5.14. Schematic representation of the problem to be modeled.

    # Mud is not cemented and all grains are free to move independently;

    # Slope sediments can have a different lithology than the quantum of erupted mud, but with coarser grain sizes at the bottom and finer grain sizes at the top of the sediment column;

Porosity for the slope sediments is unchanged during the simulation;

Porosity for a given mud current is unchanged during the flow, and different currents may have different porosity values;

Only thicknesses of the slope sediments and the height of the mud currents are changed when deposition and erosion occur;

Topography of the slope is updated after a given sediment current ends;

All grain sizes have the same grain density;

Series flows are allowed.

In MOSED3D, a discrete method, similar to Tetzlaff (1990), is used to simulate the transport, deposition and erosion of sediments during a mud flow in three dimensions. The advantages of such a discrete method are that less CPU time is required and there is more flexibility to control the simulation. The disadvantage is that the simulation is not explicitly time dependent, as in the deterministic methods of 3D simulation of mud flow (Bitzer and Pflug, 1989). As a consequence not all potential energy in the mud flow is necessarily fully converted before the sediment hits the numerical "boundary wall" for a spatially small simulation. Care must be taken to ensure that the simulation volume is large enough so that boundary wall effects are far removed from the domain of interest.

As a test case the Chirag area in the offshore Caspian Sea has been chosen. Figure 5.15a shows the bathymetric map of the area with a structural dome, corresponding to the location of a mud volcano. The broken line indicates a zone of possible eruptive centers. Figures 5.15b shows the same surface mapped onto a computer after digitizing.

The lithological content of mud, as well as of the sediments on the slopes of the volcano, is taken to consist of clay (70%); fine sand, very fine sand, coarse silt and fine silt (5% each); and from medium sand up to pebbles (2% of each fraction).

Default values of eleven modeling parameters are used for the first run. These default values are as follows:

PSB = 70% (porosity of the basin slope sediments)
PGC = 70% (porosity of the sediment current)
DG = 2650 kg/m^3 (density of the sediment matrix)
DF = 1025 kg/m^3 (density of the fluid)
VF = 0.0009 N s/m^2 (viscosity of the fluid)
DWF0 = 0.008 (Darcy-Weisbach friction coefficient for bed)
DWF1 = 0.01 (Darcy-Weisbach friction coefficient for fluid)
CON_T = 2.50 (constant in calculation of transport capacity)
CON_E = 1.20 (constant in calculation of shear stress for erosion)
HHGO = 0.50 (Erosion/deposition coefficient)
ELEFT = 0.70 (Residual energy fraction in the mud flow due to energy loss during deposition (erosion))

The eruption center was first put very close to the steepest slope, which causes a long mud tongue to develop. Figure 5.16 shows the track of the mud current movement when a mud volume of 200 m (length) x 200 m (width) x 10m (height) (= 400,000 m^3) was released. This volume is on the same order

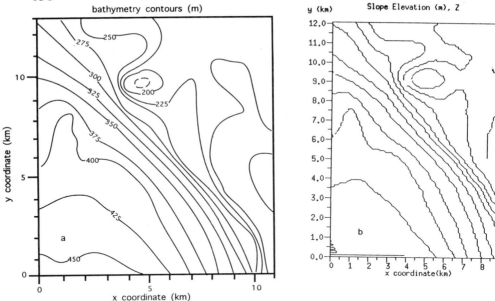

Figure 5.15. (a) Sea-bed topography (bathymetry contours in meters) over lateral and vertical scales of 10 km in the Chirag region of the offshore South Caspian Basin. The mud volcano is shown centered at coordinates 5 km laterally, and 10 km vertically; (b) digitized map of the information from figure 5.15a.

as the <u>average</u> volume of mud breccia ejected during the eruptions of observed mud volcanoes. The mud was released from the southwest region of the diapir so that the steepest topographic slope was encountered by the mud. The mud flow has a maximal extent of about 4 km but the bulk of the mud is deposited within about 2 km of the source as can be seen by the distribution of the number of tracks. The corresponding width of the mud flow is only about 300 m for the bulk of the flow, and less than 100 m at the distal part (4 km) of the flow. On the other hand, from the hazard point of view it is of greater interest to evaluate mud eruptions of maximal strength. Figure 5.17 shows the mud current movement when 200m (length) x 200m (width) x 50m (height) (= 2MMm^3) of mud was released. This volume of mud is observed during some eruptions of onshore mud volcanoes; while the probability is only 10^{-7} that ejected mud volume during a new eruption will exceed 2MMm^3 (Bagirov, Nadirov and Lerche, 1996; Section A of this Chapter), nevertheless this situation likely represents a worst case hazard. On figure 5.17 one can see that in such a case it is probable that two mud tongues of very large extent will be produced. Again the mud is released from the southwest region of the diapir, so that the maximum topographic slope is encountered. In this case note that both of the separated mud tongues are about 4 km in total length with about 500 m spacing between the tongues, each of which is about 200-400 m wide.

More interestingly, a different picture emerges when the same volume of mud is released from a broader area. Then the width of mud flow is larger,

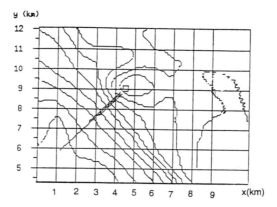

Figure 5.16. Mud current movement tracks when a moderate (0.4MMm^3) mud release occurs on the southwest side of the mud volcano for the default friction parameter HHGO set to 0.5, over a narrow area of 200 x 200 m^2.

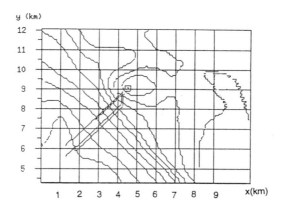

Figure 5.17. As for figure 5.16 but with a mud volume release of 2MMm^3.

corresponding more to the majority of observed cases, but not to the extreme worst case. Figures 5.18a,b,c show mud flow tracks for a 2MMm^3 release with different modeling parameters (Fig. 5.18a for HHGO=0.5; Fig. 5.18b for HHGO=0.1; and Fig. 5.18c for HHGO=0.9). One can see that the higher the value of HHGO, the shorter and wider is the mud flow tongue. The point here is that the distribution of tracks from the wider region of mud release encounter different topographic gradients and, in addition, the larger the value of HHGO the faster the mud is brought to a halt. Thus, relative to the default case of HHGO=0.5 (shown in figure 5.18a), the low value of HHGO=0.1, depicted in figure 5.18b, indicates a more uniform filling of mud across the width of the flow to about 1 km lateral scale. By way of comparison, the high value of HHGO=0.9 depicted in figure 5.18c indicates little lateral mud flow, but a concentrated core of mud flowing at the center of the turbidite. Figure 5.18b shows a more uniform mud flow extending out to about 4 km length, while the

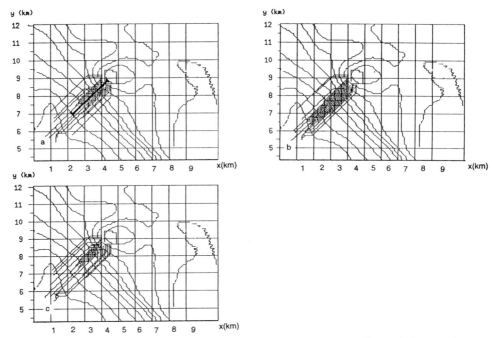

Figure 5.18. Mud current movement tracks when a 2MMm^3 mud flow volume is released on the southwest side of the mud volcano over a broad area of 400 x 400 m^2 for different values of the friction parameter, HHGO: (a) HHGO=0.5; (b) HHGO=0.1; (c) HHGO=0.9.

central core flow of figure 5.18c is limited to around 2 km length for the bulk of the released mud.

For the default value of HHGO=0.5, the topography of the sea-bed prior to mud release is shown on figure 5.19a; while the corresponding topography after release of 2MMm^3 of mud is shown on figure 5.19b, indicating the production of a broad, flat region near the diapir and a slight flattening along the mud flow. Along the cross-section marked on figure 5.18a by a bold line terminated by stubs at each end, one can also plot the change in thickness of the sediment surface (denoted by S-surface on figures 5.20a and 5.20b) both before and after the mud flow. Figure 5.20a shows the cross-section before the eruption, Fig. 5.20b after eruption, and Fig. 5.20c shows the mud flow overlain on the original sediment surface (hatchured area). From figure 5.20c one can see that the mud flow thickness has reached 20-25 m in some places. On the other hand, in spite of the very long track of the mud - up to 4 km (Fig. 5.18a) - the thickness of the flow after 2 km from the release position is negligible (Fig. 5.20c).

When the position of the eruptive center is moved to the east side of the mud volcano, then the mud flow tongue is wider, shorter and thinner (Fig. 5.21a,b) than for the default case. The reason for this shift in mud shape is that the eastern side topography is considerably flatter than the southwest side, so that there is not as much opportunity for mud to reach a high slope region before deposition terminates the flow. The corresponding cross-section (marked by a bold line terminated by stubs on figure 5.21a) after mud deposition is shown on figure 5.21b by the hatchured area overlain on

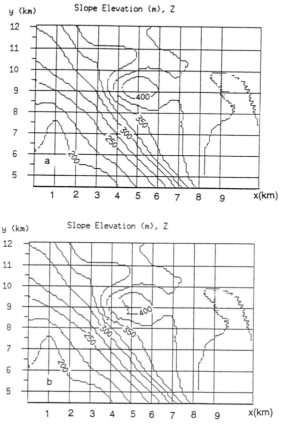

Figure 5.19. Topographic changes brought about by release of 2MMm^3 of mud over a broad area of 400 x 400 m^2: (a) topography prior to the mud release; (b) topography after the mud flow ceases.

the pre-release topography, indicating deposition only to about 1 km from the release position.

As the position of the eruptive center is moved around the diapir the overall patterns of flow change because of the topographic variations around the diapir onto which the mud flows are released. Figures 5.22a-g show eight such patterns, each for a released mud volume of 2MMm^3 and a value of 0.5 for HHGO. In general, releases from the western and southern flanks of the mud volcano travel furthest and are widest because steeper topographies are encountered by the flowing mud. Releases from the northern and eastern flanks of the mud volcano tend to encounter flatter topographic regions so that the mud flows tend to pool locally within about a kilometer or so of their release positions, with widths comparable to their lengths, representing broad, but short, flows as compared to the longer, but narrower, flows occurring to the south and west.

Because one does not know ahead of time where a mud release will occur on a mud volcano, we have taken the results of each fixed volume mud flow, released with a volume of 2MMm^3 with a value of 0.5 for HHGO, and superposed the mud flow tracks for different release positions on the mud volcano. The result was then divided by the number of release cases run to provide a region around a mud volcano where there is a significant hazard of mud flow. Shown on figure 5.23a is the overall pattern of likely hazard area around the mud volcano in the Chirag area for a mud flow thickness of 5 m or greater. Note the prevalence of the southwest region, as expected given the higher topographic slope in that area. If attention is restricted to mud flows producing a thickness of 10 m or greater, then the area of hazard around the mud volcano shrinks as shown in figure 5.23b, but there is still a south-west "tongue" due to the higher topographic slope.

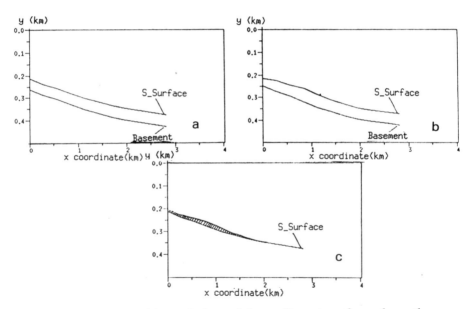

Figure 5.20. Topographic variation of the sediment surface along the cross-section marked on figure 5.18a by the stub-ended bold line: (a) topography along the section prior to mud release; (b) topography after mud flow ceases; (c) superposed topography of the mud flow thickness (hatchured region) on the original sediment surface.

D. Conclusions

The purpose of this chapter is to provide an assessment of likely hazards that could influence operational equipment in the South Caspian Basin due to potential mud flows released by mud volcanoes. A statistical data analysis shows that on land there is a 95% probability that the length of mud flow from an eruptive center is less than about 340 m, and that the mud width does not exceed 260 m; the average values of these parameters are 74 m and 56 m, respectively. For the marine environment the corresponding values have been established using a dynamical model, which is topographically sensitive. In the Chirag area, which was chosen as a test case, a zone of approximately 4x4 km^2 can be considered as a maximal hazard area, where final mud accumulation is greater than 5 meters. For 10 m mud thickness the hazard zone has about 2.5x2 km^2 size around the volcano.

Here we considered the maximal hazards. The argument for considering a worst case assessment, based on the historical record of mud flows recorded for land-based mud volcanoes, is that one should plan for a worst case hazard even if the probability of occurrence is low. One can, presumably, then accommodate for higher probability, but lower risk, hazards.

The use of a mud flow code to investigate such hazards then enables identification of not only the most likely directions of hazard for operational equipment, but also the likely mud-flow distances from the diapir (length and width) at which significant hazards could occur.

The criterion of concern is where on a mud volcano a mud release will occur. Because one does not have prior knowledge of such locations, and because land-based statistics do not indicate any distinguishable preferences for different sides of volcanoes as release conditions, it is appropriate to put together a suite of potential release sites and then consider the average as a hazard domain around a mud diapir. Depending on the mud thickness that one can gear equipment to stand up against, one then has a hazard position statement at different criteria of strength of release and frictional deposition of mud.

In this way high risk and low risk regions for siting rigs, platforms, pipelines and allied infrastructure equipment can be identified prior to a potential mud flow, which could otherwise be disastrous rather than just inconvenient. And that is the purpose for the calculations reported here.

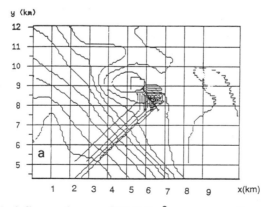

Figure 5.21a. Mud flow release of 2MMm^3 occurs on the eastern side of the mud volcano over a broad release area of 400 x 400 m^2: patterns of mud flow tracks.

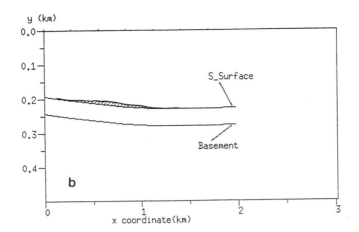

Figure 5.21b. Mud flow release of 2MMm^3 occurs on the eastern side of the mud volcano over a broad release area of 400 x 400 m^2: variation of topography along the cross-section (marked on figure 5.21a by a stub-ended bold line) caused by the mud flow (hatchured region).

Appendix A - Mathematical Formulation

MOSED3D is based on the concept that a mud current moves along a three dimensional surface according to the balance between gravity (driving force) and friction (resistance force). Changes in the velocity of the current cause changes in the transport capacity which, in large, controls the deposition and erosion of mud and sediments. The rates of deposition and erosion also depend on the basin slope and on the sediment type being carried by the current. The basic equations governing the current flow, as well as the transport, deposition and erosion of sediment and mud, are taken from Allen (1985). Some modifications are made to the equations in order to fit the needs of MOSED3D.

1. Gravity current flow

Consider a flow element as a section of the flow of unit width and fixed streamwise length. The driving force per unit width is the downslope component of the immersed weight of the element. The driving force will increase with the thickness and excess density of the current, and with the angle of the slope. The resisting force comes from friction between (i) the current and the bed, and (ii) the current and the medium. From Allen (1985) the mean flow velocity of the current is given as

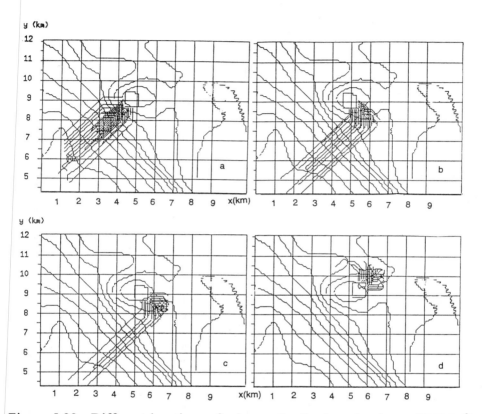

Figure 5.22. Different locations of release of a fixed mud volume (2MMm³) show different patterns of mud flow depending on surrounding topographic gradients. Figures 5.22a-d show, respectively, how the flow is altered for a southern area of release as the area is moved systematically across the mud volcano from west to east; while figures 5.22e-h show, respectively, the different flow patterns as a northern area of mud release is moved systematically across the mud volcano from east to west.

144

$$V_a = \left[\frac{8\sin\beta}{(f_0+f_1)} \frac{(\rho_2-\rho_1)}{\rho_2} gh \right]^{1/2} \quad ms^{-1} \tag{A1}$$

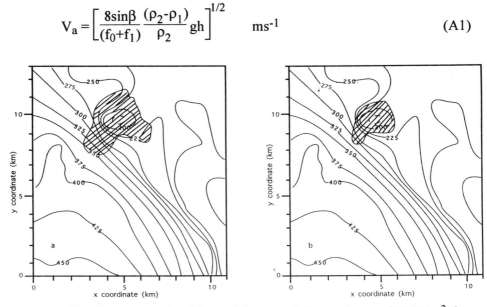

Figure 5.23. Area of maximal hazard for a mud flow release of 2MMm³ from anywhere on the mud volcano. The hatchured regions correspond to a final mud accumulation of greater than 5 m (figure 5.23a) and 10 m (figure 5.23b), respectively.

where V_a is the mean flow velocity of the current, h the thickness of the flow element, ρ_1 and ρ_2 the densities of the ambient medium (fluid) and the current respectively, g the acceleration due to gravity, β the slope angle, and f_0 and f_1 the Darcy-Weisbach friction coefficients for the bed and the medium respectively. Values of f_0 usually vary from about ~0.006 to ~0.06 for rivers, and values of f_1 are usually less than 0.01 (Middleton, 1966).

2. Mud transport

The theoretical sediment load which a current can carry is (Allen, 1985)

$$M_t = M_b + M_s = \frac{J_b}{V_b} + \frac{J_s}{V_s} \quad kgm^{-2} \tag{A2}$$

where M_t is the total theoretical mud load, M_b and M_s are the theoretical bed load and suspended load respectively, J_b and J_s are the mass transport rates for bedload and suspended load respectively, and V_b and V_s are the respective transport velocities (usually $V_b \approx 0.2\,V_a$ and $V_s \approx V_a$).

Equation (A2) shows that the total transport load consists of two components, bed load and suspended load. According to Bagnold's (1966) semi-empirical formulae, we have

$$J_t = J_b + J_s = \frac{\rho}{(\rho-\rho_1)g}(0.148+0.01\frac{V_a}{8})\frac{f_o}{V_f}\rho_2(V_a)^3 \qquad kgm^{-1}s^{-1}$$

(A3)

and
$$J_s/J_b = 0.068\frac{V_a}{V_f}$$
(A4)

where J_t is the total-load transport rate, ρ_1 the density of the ambient medium, ρ the grain density, V_f the terminal fall velocity of the mud particles in the current. All velocities must be in units of m/s so that the empirical constants are dimensionally correct. The terminal fall velocity V_f is defined as (Allen, 1985)

$$V_f = \frac{1}{18}\frac{(\rho-\rho_1)}{\eta}gD^2 \qquad ms^{-1}$$

(A5)

where ρ and ρ_1 are the density of grain and fluid, η the fluid viscosity, and D the grain diameter. The power factor 3 in equation (A3) is set as a user-defined constant (CON_T) in MOSED3D so that users can better control the transport capacity calculation; and this can be done because V^3 means $(V/(1m/s))^3$ so that the dimensions of the equation are not influenced by this change of power factor.

3. Mud deposition

Whether deposition will actually occur, with consequent losses of excess density and possibly thickness, depends on the balance between the mud load actually present in the gravity current and the mud load that can be theoretically supported by the forces due to the motion of the current. The mud can be deposited only if it is present in excess of the theoretical maximum transportable load (M_t). Hence no mud deposition will be expected if the actual load at all times is equal to or less than the theoretical value M_t (Allen, 1985).

The actual load M_a can be defined as

$$M_a = \rho_2 h \qquad kgm^{-2}$$

(A6)

where ρ_2 is the density of the current and h the height or thickness of the current. If $M_a > M_t$, mud deposition occurs and the amount of mud deposited in mass will be M_a-M_t. The coarser grain-size material will deposit first and the finer grain-size mud later.

In MOSED3D the deposited amount of mass is converted to thickness and a fractional constant (HHGO), ranging from 0.0 to 1.0, is introduced to control the actual thickness to be deposited, with HHGO = 1.0 being the total thickness.

4. Erosion

Two criteria are used to examine whether erosion will occur for the slope sediments. The first criterion is that the total theoretical load must be greater than the actual load, i.e. $M_t > M_a$, and the second criterion is that the shear stress (τ) of the current on the slope sediments must be greater than the critical shear stress (τ_c) for the sediments to be eroded, i.e. $\tau > \tau_c$.

The shear stress τ is defined as (Allen, 1985)

$$\tau = \frac{f_o}{8} \rho_2 V_a^2 \qquad \text{Nm}^{-2} \qquad (A7)$$

In MOSED3D, the power factor of 2 in equation (A7) is set as a user-defined constant (CON_E) for more control on the erosion capacity and, as for equation (A3), such can be done without influencing equation dimensions. The critical shear stress τ_c is defined as

$$\tau_c = \theta \, (\rho - \rho_1) \, gD \qquad \text{Nm}^{-2} \qquad (A8)$$

when θ is the dimensionless threshold shear stress, ρ and ρ_1 the density of grain particles and fluid, respectively, and D the grain diameter (Allen, 1985).

As in the case of deposition, the eroded amount is converted from mass to thickness with the factor HHGO. The dimensionless threshold shear stress, θ, has different values for different grain sizes. In MOSED3D, ten grain sizes are allowed and the corresponding θ values are taken from Allen (1985, Fig. 44, p. 58) as given in Table 5.1.

Table 5.1 - Lithology Parameters for Quanta Released.

Lithology Type	Grain Size (mm)	θ value	Lithology Code Description
pebble	>4.000	~0.05	L0
granule	2.000-4.000	~1.040	L1
very coarse sand	1.000-2.000	~0.040	L2
coarse sand	0.500-1.000	~0.035	L3
medium sand	0.250-0.500	~0.030	L4
fine sand	0.125-0.250	~0.040	L5
very fine sand	0.062-0.125	~0.060	L6
coarse silt	0.020-0.062	~0.15	L7
fine silt	0.004-0.02	~0.500	L8
clay	<0.004	~1.000	L9

6 Hydrate Hazards

A. General Hydrate Dissociation Conditions

The presence of clathrates (gas hydrates) is well established observationally in the offshore region of the South Caspian Basin, as well as in many other regions of the world. Hydrates of given composition can exist under particular pressure-temperature conditions. However, under the impact of neotectonic processes, those conditions can change. In that case part, or all, of the mass of hydrates can be dissociated and released. The dissociation can take place gradually, or explosively, depending on how fast the pressure drops or temperature increases. Five major variations of hydrate evolution are considered to illustrate possible patterns of behavior caused by variation of geological conditions:

    1. Variations in hydrate existence conditions due to sediment and mud redeposition. Due to earthquakes, volcanic eruptions or other processes, part of the sediments can be released from a sloping sea bottom and transported and redeposited in other places, thereby changing pressure conditions at the top of a hydrate layer.

    2. Variations in hydrate existence conditions due to glacial-interglacial conditions. Removal of ice cover not only decreases overlying pressure, but also allows water temperatures overlying the sediments to increase.

    3. Variations in hydrates due to sea-level rise and fall.

    4. Enrichment of ethane in hydrates as a consequence of varying neotectonic conditions.

    5. Evaporation and reformation of hydrates in aeolian conditions due to winter cooling and summer heating.

    Hydrates can cause an explosive hazard for exploration rigs, production platforms and pipelines, especially in deep water conditions. All of the above geological patterns of instability should, therefore, be kept in mind when potential hazards are assessed. In addition, flame initiation by dissociation is also examined as a potential hazard, as are hazards due to drill penetration, warm circulation mud, and drill-bit heating of hydrates.

    In many areas of the world the prevalence of gas hydrates at the present-day, or inferred to have been present in the recent past, is one of the more influential factors in controlling geological sedimentation and attributes allied to the geological behavior. For instance, in the Green River Basin, Wyoming, the ENRON Corporation (1993) has estimated that a total gas (possibly some in hydrate form) reserve of 5,000 TCF exists; while in the eastern United States, offshore South Carolina, there are estimates available of major hydrates present, with associated contributions to dynamical instability

(Lowrie, 1997). Or again, in the Barents Sea, Solheim and Elverhøi (1993) have argued that ice removal led to explosive decompression of metastable hydrates, as inferred from measured 'ring craters' on the seafloor. And, in the South Caspian Sea, Ginsburg et al. (1992) and Dadashev et al. (1995) have provided core measurements from three areas with a high content (up to 35%) of hydrates in recovered core, again indicative of dominance of hydrates. (Direct gravity-core sampling was done of hydrates at or near the crests of Buzdag, Elm and Abikh mud volcanoes in water depths of 480 m, 660 m, and 600 m, respectively).

The list of regions where hydrates are involved in influencing behaviors can be extended almost, it seems, indefinitely. Good references for an appreciation of the details pertaining to particular basins and hydrate influence are to be found in Kaplan (1972) and U.S. Geol. Survey Professional Paper 1570 (1993).

In nature, hydrates can form in the north Arctic zone, especially in zones of permafrost, which itself is a good seal and can form good traps for gas accumulations. So gas, slowly moving towards the surface, is stopped and accumulates at the permafrost region. Naturally the near-surface temperature is then low and hydrate crystals start to form. The thickness of the hydrate zone is defined by the geothermal gradient, or how fast the earth warms with depth.

Another zone, favorable for hydrate formation, is the sea-bottom. The pressure in the sea increases with depth by about 1 atm for every 10 m of water column. So, by a depth of 500 m the water pressure will reach more than about 50 atmospheres. At that depth the sea bottom temperature usually does not exceed 4-6°C. As a result of biological degradation of organic matter (consisting of dead sea plants and animals), the sediments at near sea bottom are saturated in carbon dioxide, hydrocarbon gases and sulfur gases. So all four components are present for hydrate formation, and in this way layers of hydrates do form in the sediments near sea-bottom. These layers are covered by a new layer of sediments, again containing organic matter, and the cycle continues. The lower boundary of the hydrate formation zone is defined by the intensity of the heat flow coming from the deep zones of the earth.

Hydrates can exist not only in deep, cold, water conditions. In the Gulf of Mexico in high temperature waters (20°C), hydrates were found (Hot Ice, 1991). An explanation is that the precise chemical composition of hydrates has a major role to play in determining the stability domain of a hydrate. For instance, ethane addition to methane allows hydrates to exist at much lower pressures and higher temperatures than for a pure methane hydrate (Baker, 1972). This tremendous stability sensitivity to the hydrate composition is important because hydrates can then exist at shallower overlying water depths and warmer sea-bottom temperatures than for pure methane hydrates. But current neotectonical processes and climate variations can change the environmental conditions, which can lead to the formation or dissociation of hydrates in the sediments.

Even 10% of ethane in the gas mixture makes a hydrate stable at 6 atm of pressure (60 meters of water column) and 6°C, while at the same temperature a 100% methane hydrate is stable only at pressures exceeding about 40 atm (400 m of water column).

For instance, on the edges of continental shelves very often one can observe massive landslides toward the deep ocean zones. These will certainly change the pressure on the hydrate formations; dissociated gases from hydrates will then escape. Such zones have been detected near the South Carolina coast. In the zone of a huge landslide (40 miles wide), seismic information shows the absence of hydrates in the cross-section, while a massive hydrate formation is located in the adjacent area (The Bermuda Triangle, 1992; USGS Professional Paper 1570, 1993).

Escaped hydrate gas may go into solution with the overlying water and so decrease the density of the water. This bubbling water presents a real hazard to ocean vessels and even helicopters, because the decreasing density of water leads to a decrease of buoyancy and so of floatation capabilities of ships. And, when released into the air, methane and ethane lower the air density making low-flying helicopters lose lift ability, as well as possibly causing methane or ethane ignition with the hot engine.

But hydrates can also dissociate explosively when the pressure is released very fast, or the temperature increases rapidly. Indeed, such an explosive dissociation of hydrates has been inferred in the Barents Sea (after recent ice removal of about 3 km thickness) by the presence of numerous large craters on the sea-floor (Solheim and Elverhøi, 1993).

The purpose of the chapter is to provide numerical orders of magnitude to describe the influence of major geological variations on hydrates, and of some of the influences of hydrates as a consequence of changed geological variations. In this way one can estimate the hazard and, in addition, one can provide recommendations of how to ameliorate the hazard from hydrates and also suggest the data needed to obtain a more accurate assessment. The dominant cause of the major differences is the extreme sensitivity of hydrate existence to the pressure-temperature regime and to the precise chemical combination of methane, ethane and, possibly, higher homologues that constitute a hydrate. The hazard arises from the fact that hydrates are only quasi-stable; if the temperature is increased at a fixed pressure, or the pressure decreased at a fixed temperature, or both, then it is easy to pass out of the stability field of hydrates, with the occurrence of explosive dissociation. As well as natural changes in the pressure and temperature regimes, an increase in temperature can also be produced by drilling into a mud diapir, thereby increasing the likelihood of triggering explosive behavior; while man-made changes in pressure are brought about by the mud weight in the borehole. Buzdag mud volcano in the Vezirov area of the South Caspian Basin has been chosen to illustrate the method of hydrate hazard assessment.

B. Hydrate Properties. Pressure-Temperature Stability Fields for Hydrates

The behavior of hydrates is summed up by the pressure-temperature curves of figure 6.1. For a pure methane/water hydrate it is required that the pressure, P, at a given temperature <u>exceed</u> the bounding curve of figure 6.1, which can be written accurately enough for the purposes here as

$$P = P_M \exp (bT) \qquad (6.1)$$

where T is the temperature in °C, b is a scaling constant which Miller (1972) quotes as 0.04 $\ln_e$ 10 ≈ 0.09 (°C^{-1}), and P_M is a scaling pressure of about 25 atmospheres at T=0°C. For a pure <u>ethane</u> hydrate the bounding curve is almost parallel to that for a methane hydrate but with the scaling pressure, P_E, of 5.2 atmospheres at 0°C, so that for existence of a pure ethane hydrate one only requires

$$P = P_E \exp (bT). \qquad (6.2)$$

which is a less stringent condition for ethane hydrate existence than for methane hydrates. Of great interest is the fact that even a small admixture of ethane lowers the scaling pressure, P_*, drastically from that for a pure methane hydrate. For instance, a 10% ethane - 90% methane hydrate has $P_* \cong 10$ atm, well below the value $P_M \approx 25$ atm for pure methane, and only about a factor 2 higher than for a pure ethane hydrate ($P_E \cong 5.2$ atm). This fact will become of great significant as we consider cyclic geological conditions. Note for future reference that 10 atm (5.2 atm) at 0°C corresponds to 100 m (52m) water, while 40 atm corresponds to 400 m water. A small addition of ethane massively increases the stability regime for an ethane-methane hydrate compared to a pure methane hydrate at the same temperature. This stability sensitivity to the hydrate composition is important in the South Caspian Basin because hydrates can then exist at shallower overlying water depths and warmer sea-bottom temperatures than for pure methane hydrates. Note, for instance, from Figure 6.1 that a hydrate composed of 10% ethane - 90% methane at a temperature of 6°C is stable at pressures in excess of about 20 atmospheres, but a 100% methane hydrate is stable only at pressures exceeding about 40 atm (20 atm corresponds to about 200 m.w.e., and 40 atm to 400 m.w.e.; Baker, 1972). For Buzdag, seven gas hydrate samples indicate a methane content from 59-87% and an ethane content of 10-19%, with higher homologues making up the balance (Ginsburg et al., 1992); while at Elm and Abikh methane concentration in hydrates is between 81-96%, with ethane and traces of higher homologues picking up the remaining 4-19%. Also, direct sampling of gas hydrate waters at the three sites indicates that it is water of mud volcano origin and not sea water that is incorporated in the hydrates (Ginsburg et al., 1992).

Hydrate concentration in the sediments cored from the Vezirov diapir varies from 5-35% at water depths of about 500 m; while in the case of the Elm hydrates of the Azizbehov structure, hydrate concentration in sediments reaches 15-20% (Ginsburg et al., 1992; Dadashev et al., 1995) in water depths of 660 m.

The stability field of the known hydrates would appear to indicate that around 200 m of water depth (at 6°C) is sufficient to provide stability of an

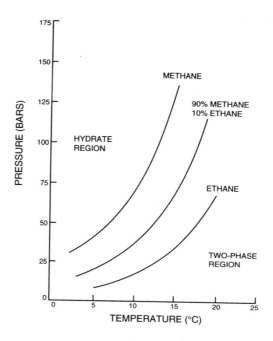

Figure 6.1. Pressure-temperature stability fields for hydrates of different composition (after Baker, 1972).

ethane/methane hydrate, that hydrates in gas-charged diapirs are likely to be a fairly common occurrence, and to provide a fairly rich (5-35%) content in surficial and near-surficial sediments.

C. Phase Diagrams and Evolution Tracks

Five major variations of hydrate evolution are considered here to illustrate possible patterns of behavior caused by variations of geological conditions. In turn, these cases are: (i) variation in hydrate existence due to sediment redeposition; (ii) variations in hydrate existence conditions due to glacial-interglacial conditions; (iii) variations in hydrates due to sea-level rise and fall; (iv) enrichment of ethane in hydrates as a consequence of varying geological conditions; and (v) evaporation and re-formation of hydrates in aeolian conditions due to winter cooling and summer heating.

1. Evolutionary tracks under sediment deposition-redeposition conditions

How deep in the sediments hydrate of a given composition can exist depends on the values of temperature gradient and the density of the sediments. For a water depth, h_w, a bottom water temperature T_b, and a sediment density ρ_s, the pressure P at a distance z into the sediments is

$$P = g\rho_w h_w + g\rho_s z \qquad (6.3)$$

where ρ_w is the water density. The corresponding temperature at a distance z into the sediments is

$$T = T_b + Gz \qquad (6.4)$$

where G is the vertical thermal gradient.

So in the deeper sediments, the pressure increases but temperature increases too. To estimate the maximal depth from the sea bottom, z_*, where the hydrates of a given composition can exist, one then has the relation

$$\log_{10}((\rho_w g h_w + \rho_s g z_*)/P_0) \geq 0.04 \, (T_b + Gz_*) \qquad (6.5)$$

where P_0 is the critical pressure at 0°C. Now P_0 depends sensitively on the composition of the gases which formed the clathrates. As sediments are loaded on the sea-bottom the hydrates, existing in the near surface sediments, become buried deeper which will increase the pressure on the hydrates. But, with time, the newly deposited sediments will be heated, so the temperature will increase. If the sedimentation is slow enough then there will be time for the sediments to be heated such that the temperature gradient remains constant. In such a case the maximal depth, z_*, where hydrates can exist is as above. In that case the existence time of hydrates is z_*/v, where v is the rate of sedimentation. An evolutionary track of the hydrates is shown on fig. 6.2a. Initially the pressure-temperature conditions where the hydrates exist are described by point A, which corresponds to the sea-bottom temperature and hydrostatic pressure. With continued sedimentation both the pressure and temperature increase as shown on fig. 6.2a. Point A' corresponds to the critical point, after which dissociation of the hydrates occurs.

If the sedimentation rate is high then the deposited sediments will remain cool and only slowly warm up by the heat flow coming from the deep basinal zones. Such rapid sedimentation can be observed in a number of basins. For instance, in the South Caspian Basin about 10 km thickness of sediments was deposited during the last 5 million years (Lerche et al., 1996). The rates of sedimentation were much higher than the rates of heating, which caused the very low observed shallow thermal gradients of 1.8-2.0°C/100 m.

Another case is that of mud flows or landslides in the South Caspian Basin in particular. During the eruptions of mud volcanoes, occurring in many places in the world, the mud flows can form covers of tens of meters in thickness. That load of sediment and mud mixture occurs during a few hours or days, and can be considered almost instantaneous on the geological time scale. Certainly, the temperature of that load will be the same as sea-bottom temperature, or even lower because of adiabatic degasification which always take place during eruptions of mud volcanoes. In such a case initially one has an increasing pressure at almost constant temperature and then a slowly increasing temperature at an almost constant (higher) pressure. Hydrates are first moved to more stable conditions and, with temperature then increasing, they then move to more and more unstable conditions, eventually crossing the threshold of stability (Fig. 6.2b). The same figure illustrates the case where a gas/water mixture can reach the zone of hydrate stability, with the formation of hydrates, and then is followed by their dissociation as shown on fig. 6.2b. Note that the evolution track can also stay in the zone of instability.

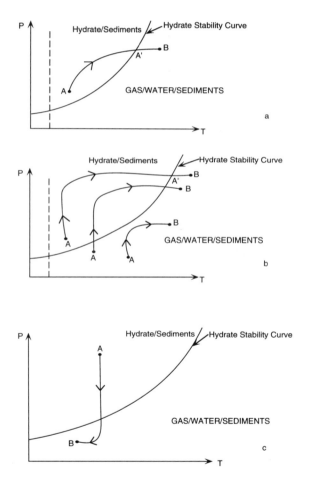

Figure 6.2. Evolutionary tracks of the gas-water mixture under sediments loading/erosion conditions. a) slow deposition; b) fast deposition; c) erosion.

2. *Evolutionary tracks under glacial-interglacial conditions*

Three major effects occur as ice loading on a sedimentary basin increases during glaciation. First, increasing ice thickness adds load to the sedimentary column, thereby increasing pressure on the underlying sediments (in the Barents Sea glacial thicknesses estimates are estimated to be in the several km range, much in excess of the current 200-400 m mean water depth) and thus increasing hydrate stability. Second, the base of the ice is at around 0-2°C, so that the temperature in underlying sediments is also lowered, thereby increasing the stability conditions for hydrates. Third, ice has a very low permeability so that gases generated in the sediments will be trapped to a greater extent with an ice overlay above the sediments compared to only a water overlay. This trapping increases sub-ice methane and ethane concentrations leading to a greater likelihood of hydrate formation. By way of contrast, removal of ice cover decreases overlying pressure, allows water temperatures overlying the sediments to be increased, and so the only support for containment of any gas is the pre-existing hydrate cover. If ice pressure removal is high enough and/or if the water temperature increases enough then hydrates will explosively dissociate, and gas underlying the hydrates will then no longer be so easily contained. As mentioned already, explosive craters observed on the sea-bottom floor of the Barents Sea (Solheim and Elverhøi, 1993) have been ascribed to this hydrate dissociation. An evolutionary track of a hydrate is sketched on a pressure-temperature phase diagram in figure 6.3 during a cycle from zero ice cover through to maximum ice cover and then back to zero ice cover. Initially, gas in near surface sediments (either as free-phase gas in the pore space or in water solution in the pores) is exhibited at point A of figure 6.3. As ice builds up the pressure increases in direct proportion to the overlying ice-thickness but, because of the finite thermal conductivity, there is a delay in the sediment temperature evolving contemporaneously with the increase in ice cover. As the ice cover increases and the sediment temperature starts to cool, eventually the sediment conditions cross the pressure-temperature curve for hydrates at the point A' of figure 6.3, so that the gas then goes into hydrate formation with available pore water. As ice thickness increases further and sediment temperature is lowered even more, the hydrate traces out an evolutionary track along the phase diagram of figure 6.3 towards the point B, representing highest pressure and lowest temperature at maximum ice cover. Once the ice is removed the pressure due to ice drops precipitously, and the hydrate moves from point B to point C in the phase diagram. If sediments are not eroded by ice-scour at the time of ice removal, then point C lies always above the hydrate stability curve. (However, if sediments are also scoured then point C can drop below the stability curve, as sketched by the dashed line to point C' in figure 6.3, when explosive dissociation of residual hydrates can occur). The sediment temperature will slowly recover as the last ice is removed so that the hydrate will then move along an evolutionary curve from point C to point D where it will cross the hydrate stability curve and dissociate. (However, if any replacement water in the basin is not of high enough temperature, then the sedimentary increase in temperature will be low and the hydrates can cease their evolutionary track at point D'; and if the replacement water depth is deep then the pressure again increases, bottom water temperature rises to only around 4°C when the hydrates end their evolution at position D" on figure 6.3). Depending on where the hydrates end this cycle of glacial-interglacial evolution then determines their further evolution during the next cycle plus, of course, any newly generated hydrate evolution tracks created by a further supply of gas to the system, or of residual gas underlying the prior hydrate cap which did not itself form a hydrate but was trapped and sealed under the hydrate.

We have traced the hydrate evolution schematically only through one glacial-interglacial cycle because evolution through a second cycle requires a more precise

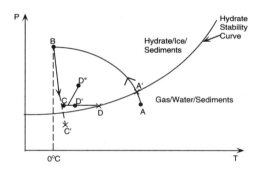

Legend:

A: Start-point of gas in water-charged sediments.

A': Pressure-temperature point of initial hydrate formation.

B: Maximum ice cover.

C: Ice removal (temperature at sediment surfce is at 0°C until final ice removal).

D: Hydrate dissociation point due to temperature increase after ice removal.

D': Hydrate evolution end if temperature increase is not high after ice removal.

D": Hydrate evolution end if water depth is high and seabottom temperature low after ice removal.

Figure 6.3. Evolutionary tracks of the gas/water mixture under glacial-interglacial conditions.

knowledge of the specific conditions of the evolutionary point of the hydrate at the end of the first cycle and, as individual and specific conditions change in a basin, so too does the evolutionary end point of the hydrate, making a tracing of the hydrate evolution through a second glacial-interglacial cycle sensitively dependent on precisely where (point C', D, D' or D") the hydrate ended up at the termination of the first cycle.

3. Evolutionary tracks under sea-level rise and/or fall

Under conditions where the water depth overlying the sediments rises and/or falls, the water pressure changes in direct proportion to the rise and fall. The stability of a hydrate then rests on two factors: (a) the magnitude of the rise and/or fall; (b) the change in temperature at the sea-bottom, which then influences how rapidly the temperature changes in the sub-sea sediments. An initially deep water depth will, in general, have a low base temperature so that both the overlying water pressure is high and the temperature low, precisely the conditions required for maximizing hydrate production. As the water depth is lowered the pressure decreases and, in shallow waters, the base water temperature also rises. Thus the stability regime for hydrates is lowered to minimum stability conditions. On the other hand, an initially shallow water depth with relatively high base temperature may be such that hydrates are not then produced.

However, as the water depth increases and, correspondingly, the base water temperature is lowered then, eventually, hydrate stability is achieved. These two situations are sketched in figures 6.4a and 6.4b, respectively. Thus, figure 6.4a shows the phase path of a hydrate during sea level fall starting at point A. The end-point of the path can be at points B, B' or B" depending on how much the sea-level falls. A drop in sea-level which is small will keep the pressure above the hydrate stability curve, thereby ending the phase path at point B, while a larger drop in sea-level will cause the phase path to cross the hydrate stability curve at point B', and end as a gas/water mix at point B". By reversing the argument, as sketched in figure 6.4b, a rise in sea-level, if small, will keep an initial (point A) gas/water mix as a gas/water mix so that the phase path will end at point B of figure 6.4b, while a larger increase in sea-level rise will allow hydrate formation when the phase path crosses point B' of figure 6.4b, and the evolution ends at point B" representing hydrate rather than a gas/water mix.

A cyclic rise and fall of sea level produces the three possible phase loops sketched in figure 6.4c, and labeled A, B and C, respectively. Because there is a thermal conductivity to sediments, there is a time-delay in sediment temperature adjusting to a changing sea-bottom temperature but little to no delay in pressure accommodation. Thus, following any one of the phase paths anti-clockwise around its loop starting at location 1 of each loop, means that an hysteresis effect occurs with slower temperature response. In loop A of figure 6.4c the rise and fall of sea-level crosses the hydrate stability line so one progresses from a gas/water mix to hydrate and back to a gas/water mix as sea level first rises (locations 1 to 2), then sediment temperature cools (locations 2 to 3), then sea-level drops (locations 2 to 4), with a final increase in sediment temperature (locations 4 to 1). In loops B and C either hydrates are preserved (loop B) or a gas/water mixture always is present (loop C). As we will see later, it is loop A types of behavior which can lead to ethane enhancement in hydrates.

Note that every time a loop A situation arises there is the potential for explosive failure of hydrates during drops in sea-level (locations 3 to 4).

4. Ethane enrichment in hydrates

In hydrates from the South Caspian Sea the observations of a high ethane/methane ratio (of order 10-20% or more) are, at first, somewhat puzzling because natural gas production rarely has more than a couple of percent "wet gas" (ethane). Pyrolysis/GCMS experiments also indicate that gas production from organic matter is massively dominated by methane not ethane. And yet, by way of contrast, the hydrates of the South Caspian Basin show a high ethane enrichment. One can, of course, always argue that that is the way the original hydrates were made and so appeal to anomalous TOC (Total Organic Carbon) or anomalous migration of ethane to produce the ethane-enriched hydrates. But the three cores with such high ethane hydrates are from three very different regions of the South Caspian Basin, suggesting that a mechanism is called for which is more pre-disposed to being generic in character. One such possibility is to invoke the arrhythmic rise and fall of the sea-level in the South Caspian Basin in association with the changes in phase stability conditions that a small addition of ethane makes to a hydrate.

The phase-diagrams of figure 6.1 exhibit the hydrate stability curves for methane hydrates, for 90% methane - 10% ethane hydrates, and for pure ethane hydrates, respectively (Baker, 1972). Note how the addition of even just 10% ethane increases the stability of a hydrate against pressure changes relative to a pure methane hydrate. So one can then visualize how variations in pressure and/or temperature can cause enrichment of ethane in hydrates. Imagine that one started with a hydrate that, on average, was dominantly methane-based with only a very small percentage ethane. Lowering pressure will, on average, cause dissociation of the hydrate. Only those spatial regions of hydrate relatively enriched (compared to the average) in ethane will survive better than regions more nearly 100% methane hydrate. Thus fluctuations in time in both pressure and temperature will systematically remove pure methane hydrates in favor of hydrates which become progressively more enriched in ethane - a survival of the fittest scenario. The only requirement is that fluctuations in hydrate composition are present to act as "seeds" for the evolutionary enhancement, and such factors are always present. Thus the ethane enrichment observed in recovered cores containing hydrates in the South Caspian Basin may be an artifact of initial hydrate production, but can also be accounted for by progressive evolution of the hydrate under changing pressure and temperature conditions after hydrate initiation - or both, of course.

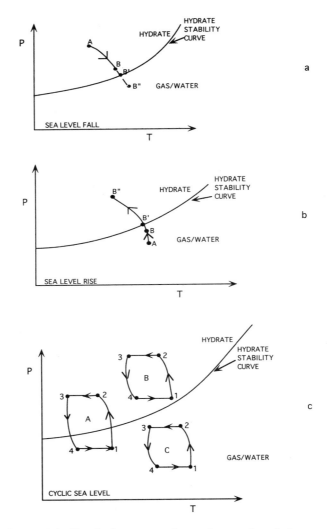

Figure 6.4. Evolutionary tracks under sea-level changes.

5. *Aeolian effects: winter/summer and hydrate evaporation/reformation*

Not all hydrates occur in marine settings. Perhaps the archtypical example of hydrates in an aeolian setting is the giant West Siberian gas field (Kaplan, 1972). The phase-diagram of figure 6.1 can be used to estimate <u>minimum</u> temperatures for aeolian hydrate occurrence as follows.

The hydrate curves of figure 6.1 for different compositions are almost parallel below about 15°C and can be written generically in the form

$$P = P_0 \exp (bT) \tag{6.6}$$

where P is pressure (in atm), T is temperature (°C), P_0 is the hydrate stability pressure at 0°C and ranges from about 40 atm for 100% methane hydrate to 5.2 atm for 100% ethane hydrate, and is 10 atm for a 90% methane - 10% ethane hydrate; here b is a scaling constant measuring the rate of curvature and has a value of about $b \cong 0.09$ °C^{-1}. Consider then that hydrates could exist in an aeolian environment at the sedimentary surface, so that the pressure is that of the overlying column of air, P = 1 atm. The corresponding maximum temperature at which a hydrate could then exist is

$$T_{crit} = b^{-1} \ln (P/P_0) \text{ °C.} \tag{6.7}$$

For 100% methane hydrate one has $T_{crit} \cong$ -41°C; for 90% methane - 10% ethane hydrate one has $T_{crit} \cong$ -26°C; while for 100% ethane hydrate one has $T_{crit} \cong$ -18°C. Thus the surface temperature must be more negative than T_{crit} to allow surface hydrate formation. The major influence of any sedimentary cover is clear: the sediment pressure is about 1 atm for every 5 m of sediment so that if one were to consider even 5 m depth into the sediments (P $\cong$ 2 atm) then the corresponding critical temperatures are $T_{crit} \cong$ -33°C (100% methane), $T_{crit} \cong$ -18°C (90% methane-10% ethane), and $T_{crit} \cong$ -11°C (100% ethane). There is an extremely rapid increase in the maximum critical temperature for hydrate occurrence with increasing depth into the sediments.

Now consider the effects of variations in surface temperature caused by winter-summer variations. These temperature variations, when carried into the subsurface, provide a spatially periodic amplitude variation in temperature which declines with increasing penetration depth and which varies in time in concert with the surface temperature. Thus at any pre-chosen depth the temperature ranges periodically between a minimum and a maximum, with the amplitude becoming smaller at greater depth. If the temperature fluctuations at a given depth cross the critical temperature then, on the high temperature side the hydrates sublime and can reform on the low temperature side, provided gas does not have time to escape the sediments or if more gas is available. A sketch of the patterns of behavior envisioned is given in figure 6.5. The reason for the hysteresis path from summer to winter and return to summer is due to the fact that sublimation of the hydrate will decrease the load as gas escapes so that there is always some hysteresis, although that declines with increasing depth as shown with curve A (higher pressure) compared to curve B.

It is also possible that the fluctuations in surface temperature when carried to depth are not sufficient to allow the maximum temperature to cross the hydrate stability curve; this sort of situation is depicted by curve C on figure 6.5. In this case hydrate preservation is maintained. Again, because of the difference in the hydrate stability curves for different compositions, it is easier for an annual temperature wave to cross the methane hydrate stability curve

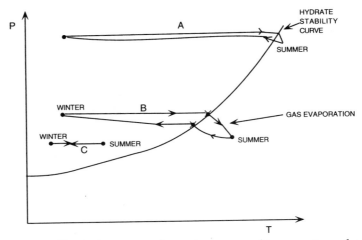

Figure 6.5. Evolutionary tracks under seasonal temperature changes.

than the ethane hydrate stability curve - again leading to a loss of methane relative to ethane and so a progressive ethane hydrate enrichment over many annual cycles of winter to summer conditions.

D. Hazard Factors for Hydrates

Three factors are of importance in determining the dissociation of hydrates. First is the temperature gradient in the sediments; second is the composition of the hydrate, which determines the critical pressure and temperature limits for hydrate existence; third is the total sediment volume actually occupied by hydrates.

In the case of the offshore South Caspian Basin, regional temperature gradients would seem to be about 1.8-2°C/100 m, at least as deep as the Productive Series of the Plio-Pleistocene (Tagiyev et al., 1996), while undisturbed water bottom temperatures are around 4-6°C. However, the presence of gas in mud diapirs is known to lower the temperature in the crestal domain of a diapir and local surrounds by up to 30-50°C relative to regional measurements at the same sediment depth (Bagirov and Lerche, 1996). For a mud diapir of typical diameter 10 km, this lowered temperature corresponds to a lowering of the vertical temperature gradient by about 0.3-0.5°C/100 m, giving a total temperature gradient in and near a diapir of around 1.3-1.7°C/100 m.

For a 10% ethane - 90% methane hydrate the critical pressure for hydrate production is lowered from the 26 atm needed for a pure methane hydrate at 0°C to about 10 atm (Baker, 1972) at 0°C and the hydrate stability region is increased (Figure 6.1). For a water depth, h_w, a bottom water temperature T_b, and a sediment density ρ_s, the pressure P at a distance z into the sediments is

$$P = g\rho_w h_w + g\rho_s z \qquad (6.8)$$

where ρ_w is the water density. The corresponding temperature at a distance z into the sediments is

$$T = T_b + Gz \qquad (6.9)$$

where G is the vertical thermal gradient. Because the stability curves are almost parallel for methane hydrate and a hydrate composed of 10% ethane - 90% methane (Baker, 1972), one can write the pressure-temperature stability relation based on the methane curve as

$$\log_{10}(P/P_o) = 0.04\ T_c + 3 \times 10^{-4}\ T_c^2 \qquad (6.10)$$

where P_o is the critical pressure at 0°C and T_c is the temperature in °C (Miller, 1972).

To estimate the total sediment thickness, z_*, that could maximally contain a 10% ethane - 90% methane hydrate, one then has the relation

$$\log_{10}((\rho_w g h_w + \rho_s g z_*)/P_o) =$$
$$0.04(T_b + Gz_*) + 3 \times 10^{-4}(T_b + Gz_*)^2 \qquad (6.11)$$

where T_b is the bottom-water temperature in °C.

For instance, if $z_* = 0$, then equation (6.11) requires

$$h_w \geq P_o(\rho_w g)^{-1} \exp\{(\ln_e 10)\ [0.04 T_b + 3 \times 10^{-4}\ T_b^2]\} \qquad (6.12)$$

which provides a minimum water depth to support hydrate occurrence. For P_o = 10 atm, T_b = 6°C, the minimum water depth is $h_{w,min}$ =210 m; while for T_b = 4°C and P_o = 10 atm, the minimum water depth is reduced to $h_{w,min}$ = 170 m. As the ethane concentration increases so the critical pressure P_o at 0°C is lowered below 10 atm, until, at 100% ethane, the critical pressure is 5.2 atm (Miller, 1972) corresponding to a minimum water depth of about 85 m at 4°C. Once the water depth is greater than the minimum needed for hydrate production, it then becomes a simple matter to use inequality (6.11) to estimate the maximum sediment thickness capable of sustaining a hydrate.

Consider for illustration the Vezirov diapir in 500 m of water, where the water-bottom temperature is 4°C. We carry through the calculations for a 10% ethane - 90% methane hydrate (P_o = 10 atm at 0°C), and with $\rho_s \cong 1.6$ gcm^{-3}, corresponding to near surficial unconsolidated sediments. Then for hydrate existence one requires

$$\log_{10} [5 + 1.6 \, z_{100}] = 0.16 + 0.055 z_{100} + 5.10^{-4} \, z_{100}^2 \qquad (6.13)$$

where z_{100} is the critical thickness z_* in units of 100 m.

Inequality (6.13) is satisfied as long as z_{100} is less than about 5.2, corresponding to a hydrate potential thickness of $z_* = 520$m, at which depth into the sediments the temperature is about $(4 + 1.3\times5.2)°C = 11°C$, and the dissociation pressure is about 130 atm. From an explosive decompression viewpoint it then would seem that a mud-diapir of radius R, filled to a volume fraction f with hydrates to a depth z_*, has an explosive volumetric potential, V, of

$$V \cong \pi \, R^2 z_* \, f. \qquad (6.14)$$

For a 20% volume fraction (f) of hydrate (10% ethane-90% methane) in a 5 km radius (R) mud diapir, with the hydrate extending to its maximum sediment depth (z_*) of about 500 m, the total hydrate volume is $V \cong 8\times10^9 m^3$.

Because the volume V is roughly 1/7 hydrate gases, a total gas volume of $V_{gas} \cong 10^9 m^3$ is available.

Due to the de Laval effect through narrow pore throats, Ivanov and Guliev (1987) calculate the corresponding water/gas temperature rise at some 280-560°K. Thus the explosive mix would be at a temperature of between about 290-570°C. Given that the air/methane self-ignition temperature is about 537°C (Guliev, 1996; Ivanov and Guliev, 1987) it follows that: (a) water would convert to steam; (b) the explosive pressure of 130 atm is compensated for by only 50 atm of overlying water pressure, so that 80 atm of dynamic pressure is available to blow a hole in the South Caspian ocean, releasing gas from the hydrates to combine with air and self-ignite; (c) a quantity of super-heated steam will also be produced.

E. Seismic Effects and Mud Flows

The purpose of performing these calculations has been to provide estimates and ranges for potential hazards from hydrates in the South Caspian Basin. The problem of addressing every nuance of every possible situation is avoided by using empirical measurements to provide ranges of parameters that can influence hydrate dissociation. In that way one can estimate a likely worst case scenario.

It would appear that hydrates can occur in as shallow as 50 m of water, if the hydrate composition is dominated by ethane, while 150-250 m of water is needed at a 90% methane composition.

It would appear that stability of hydrates can occupy as much as about 500 m of sediments when overlain by about 400-500 m of water. The cooling of the crestal regions of a mud diapir by the low thermal conductivity of gas-charged sediments is partly responsible for this large thickness of potential hydrates. The corresponding dissociation pressure is about 130 atm, quite capable of blasting a hole through the overlying 500 m of water, which only provides a retaining pressure of 50 atm, leaving a residual 80 atm of dynamic drive pressure. Note that if the 80 atmospheres of pressure is all converted to a dynamic drive on the mud, then a column of mud to a height of about 800 m can be produced. Thus, on the order of 1 km (or a little less) could be extruded from the mud diapir at a maximum speed of around 0.01-0.1 km/sec, so that a hazard event would take on the order of $10-10^2$ sec to become established. This speed of hazard development is so fast as to require constant monitoring for its actualization probability. Energy at dissociation is sufficiently large to produce temperatures of 290-570°C (Ivanov and Guliev, 1987), enough to convert seawater to super-heated steam, and enough for self-ignition of released methane in contact with air.

The identification by seismic methods of the top of a hydrate layer in sediments can be used directly to estimate the temperature gradient in the sediments as follows. The seismic information is depth-converted from two-way travel-time to physical depth using a velocity analysis code, so that both the depth of overlying water, h_w, and the depth, D, into the sediments to the top of the detected hydrate reflector are known. Then the total pressure to the top of the hydrate reflector can be estimated. Hence one can use equation (6.12) in reverse to determine the temperature gradient given some assessment of likely composition of the hydrate (so that the critical pressure P_0 at 0°C can be estimated), and an estimate of water-bottom temperature. Thus, directly from seismic one can estimate the temperature gradient in a hydrate-bearing mud diapir, and so one can then estimate the maximum depth into the diapir at which hydrates could exist. In this way seismic information alone can provide an estimate of a worst-case hazard from hydrate dissociation.

As noted by Claypool and Kaplan (1972), sediments have seismic velocities of 2.2-2.5 km/sec under conditions that support hydrate formation, compared to normal velocities of 1.7-1.9 km/sec for similar sediments without hydrates. This increase in seismic velocity can be used as a diagnostic of the thickness of hydrates, an increase in seismic velocity at shallow sediment depth with a deeper decrease to a more normal seismic velocity trend is indicative of a hydrate zone. The seismic reflection amplitude off the top of a hydrate should be about 6-12%. Thus, as well as being used to estimate temperature gradients in the presence of hydrates, seismic reflection information can also provide an indication of hydrate thickness.

Combining the two procedures, each of which provides a depth limit to hydrate occurrence, permits an estimate to be made of hydrate composition because, while the increased seismic velocity is relatively insensitive to hydrate composition, the temperature stability is sensitive (Figure 6.1). Thus if one can estimate the base of the hydrate zone from seismic velocity information, and because one can also estimate the temperature at the hydrate base position, as well as the total sediment plus overlying water pressure to the hydrate base, then one can use the pressure-temperature stability diagrams of Figure 6.1 to estimate the hydrate composition. A quantitative method is given in the Appendix.

Dynamic motion of a mud diapir can also lead to explosive dissociation of a hydrate-bearing mud diapir. Shown on figure 6.6 is a schematic of a diapir both before and after slippage and rotation along a decollement surface, showing how a hydrate can rise above the stability line, leading to explosive behavior. Of even greater concern is the fact that submarine mud diapirs are known to exhibit flows down their flanks, thereby changing their topography. Internal sediment failure of the unconsolidated mud can then lead to pressure and temperature changes within the sediments, which may inhibit or enhance hydrate dissolution. Catastrophic hydrate hazard associated with the reshaping of a mud diapir would seem to be a problem of pressing concern, but one which is difficult to evaluate from the viewpoints of either time-of-occurrence, or interval-between-occurrences.

Buoyancy effects of hydrates on unconsolidated sediments are also relevant. Hitchon (1972) provides estimates of hydrate densities of $\rho_H = 0.8$-0.9 gcm^{-3}. For a sediment, of density ρ_s, containing a disseminated fraction, f, of hydrates, the bulk density is $\rho_B = \rho_s - f(\rho_s-\rho_H)$; if this density is less than that of the overlying seawater, of density ρ_w, then the mix of hydrates and sediments is buoyant. The critical fraction of hydrate admixture necessary to achieve buoyancy is then $f = (\rho_s-\rho_w)/(\rho_s-\rho_H)$.

For $\rho_s \sim 1.6$ gcm^{-3}, $\rho_w \sim 1.03$ gcm^{-3}, and $\rho_H \cong 0.8$ gcm^{-3}, then f = 70%. The South Caspian measurements (to about 1-1.2 m depth into the sediments) (Ginsburg et al., 1992) indicate a fraction of only up to 35% hydrate, so that the observed sediment-hydrate mixes are not buoyant. (Alternatively: for a hydrate fraction of 35% the sediment density, ρ_s, should be less than 1.15 gcm^{-3} if buoyancy is to be achieved).

However, Ginsburg et al. (1992) also report finding crystals of hydrate (5 cm x 2 cm area, 2-4 mm thick) sitting on the ocean floor in 500 m of water. And these crystals are clearly buoyant relative to overlying sea-water. At a water-depth of h_w the buoyancy pressure is

$$P_{buoy} \cong g(\rho_w - \rho_H) h_w$$

which, with $h_w \cong 500$ m, has the numerical value $P_{buoy} \cong 10$ atm. Because the hydrate crystals are buoyant there must then be a binding affinity of hydrates to the sediments which exceeds 10 atm. When sediments are disturbed by mud flows, volcanism, etc., it is interesting to speculate that if the binding affinity is removed, then hydrate buoyancy can lead to a rapid rise of hydrate crystals through the overlying water until the critical pressure depth is reached (about 100-200 m depth of water depending on hydrate composition), when hydrate crystals will dissociate releasing methane and ethane gas bubbles which continue their rise to the ocean surface. Thus, during times of diapiric slumping, volcanic activity or submarine mud flows, one anticipates an increase in bubbling of methane and ethane gases at the ocean surface from hydrate dissociation, a strong surficial indicator of enhanced submarine activity associated with mud diapir evolution.

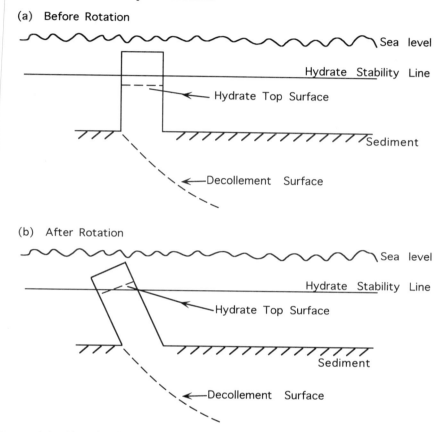

Figure 6.6. Sketch of the influence of rotation and/or slumping of a mud diapir on hydrate positioning relative to the hydrate stability line: (a) before rotation; (b) after rotation.

F. Conclusions

Hydrates should not be considered only as an interesting phenomenon which can be a future energy resource. Hazards arise from the fact that hydrates are only quasi-stable; if the temperature is increased at a fixed pressure, or the pressure decreased at a fixed temperature, or both, then it is easy to pass out of the stability field of hydrates. The tremendous destructive power of dissociated hydrates was evaluated by Ivanov and Guliev (1987) and later by Lerche and Bagirov (1993/94). And those assessments correspond to the observations: explosions on the Barents Sea floor (Solheim and Elverhøi, 1993); flaming eruptions in the Caspian Sea (Dadashev et al., 1995), when the flames of burning methane came from the bottom of the sea in water depths of hundreds of meters, and the flames reached hundreds of meters in height. During planning for hydrocarbon exploration and exploitation, one must keep in mind the possibility of such hazards. Hydrates, because of their metastable existence, can destroy platforms and pipelines; and the flames caused by hydrate explosions can present a major heating hazard to people and equipment within a radius of 10 or more kilometers. Even non-explosive dissociation of hydrates can lead to outgassing of the seawater, which can cause catastrophic consequences to vessels and floating platforms. But the present-day stable or metastable environmental conditions for hydrate existence does not mean that such conditions will stay the same or were always the same. Modern geological and neotectonical processes can easily change those conditions. The purpose of this section of the chapter was to track such changes, which can be caused by different geological, hydrological, environmental, and seasonal variations, and which can lead to massive dissociation of hydrates.

But man-made triggering of the hydrate hazard must also be considered. For instance, drilling into a mud diapir will raise the mud/hydrate temperature either by direct drill-bit heating or by warm circulating mud heating. In either event, heating hydrate-bearing mud much above about 11°C will dissociate the hydrates, leading to explosive release of pressure, and to possible rig annihilation.

However, each basin and each region of a basin require individual searches of the possible changes in the environment for hydrate formation and loss. The processes which have been considered in this chapter rarely occur separately and mutually influence each other and hydrate existence conditions.

A good example is the South Caspian area. Massive sedimentation during the last 5 million years was a main reason of extremely high overpressures, low temperature gradients and mud diapirs, the latter being bodies highly saturated in gas. Low thermal gradients allow hydrate formation to deep zones in the sediments (a few hundred meters). Sea level fluctuations bring about the formation/dissociation of hydrates, which enriches the hydrates in ethane and allows them to exist in shallower water conditions. At the same time mud volcanoes associated with the diapirs periodically erupt. Those eruptions sometimes lead to explosive dissociation, and the mud flows transport some of the sediments down the slopes of the volcanoes. Those redeposition conditions can be followed by changes of the current conditions of hydrates described in this chapter. Other regions have their own specific characteristics and should be studied individually.

Accordingly, this hazard is, perhaps, one that should be of serious concern to oil companies drilling in offshore regions of the South Caspian Basin.

Appendix - A Quantitative Procedure for Hydrate Composition Determination From Seismic Data

Suppose the water depth, h_w, the physical depth to the top of the hydrate reflector, z_1, and the depth, $z_2(>z_1)$, to the base of the hydrate reflector are all inferred from seismic velocity information. Then the pressures $P_w = gh_w\rho_w$, P_1

$= P_w + (z_1 - h_w)g\rho_s$, and $P_2 = P_1 + g\rho_s(z_2 - z_1)$ are estimated ,as is the bottom-water temperature T_b. The hydrate stability equation can be written in the generic form

$$\ln P - \ln P_o = bT + cT^2 \tag{A1}$$

where b and c are fixed constants, and P_o is hydrate composition specific (Figure 6.1).

Then let the temperature at the hydrate top be T_1 and at its base be T_2 with geothermal gradient G, so that $T_1 = T_b + G(z_1-h_w)$, $T_2 = T_1 + G(z_2-z_1) = T_1 + GH$, where H is the measured hydrate thickness.

Then

$$\ln(P_2/P_1) = G(bH + c\,(2T_b + 2G(z_1 - h_w) + HG)) \tag{A2}$$

Equation (A2) can be solved for the geothermal gradient G, which is the only unknown.

Then

$$\ln P_o = \ln P_2 - bT_2 - cT_2^2 \tag{A3}$$

so that the critical pressure P_o at 0°C is also determined. Hence from figure 6.1 one can find out which hydrate composition curve corresponds to the value P_o, and so the hydrate composition is found.

⁷Gas Hazards

One of the significant hazards, which can potentially "kill" a well during the drilling process, is the presence of gas pockets or gas accumulations. Gas changing from solution phase to free phase has a substantial energy. Overpressure, created by gas, can easily block the drilling bit and destroy the borehole. On the other hand, the gas itself is very dangerous for platform personnel and for the environment. Sulfur gas and carbon dioxide in high concentration can poison people, and hydrocarbon gases, while not so toxic, can flame.

The sizes and shapes of gas-containing geological bodies vary from relatively small, isolated lenses to huge mud diapirs. The latter are more important because they do not just present a hazard, but also are very attractive from the commercial gas production point of view.

The purpose of this chapter is to show how huge mud intrusions form and develop, as well as to estimate the concentration of gas saturated in the mud of the diapirs.

Mud diapirs occur in a number of modern accretionary wedges (Brown and Westbrook, 1988) and have significant impacts on structural development, on mass motion and heat transportation, on fluid pressure development, and on hydrocarbon generation, migration and accumulation. The association of mud volcanoes with oil and gas fields, and the activity of mud volcanoes, have had a long history of investigation by scientists who have studied the problem of the interaction between hydrocarbon generation, migration and mud volcano activity (Gubkin, 1934; Kovalevskiy, 1940; Abramovich, 1959; Dadashev and Mekhtiyev, 1975; Kastrulin, 1979; Yakubov et al., 1980; Kropotkin and Valyayev, 1981; Rakhmanov, 1987). Mud has a low permeability and so acts as a seal for hydrocarbon accumulations, with traps on the flanks and crests of diapir structures. Evolving diapirs form anticlinal structures, which provide excellent traps. Stress and strain produced in the formations surrounding a diapir during mud flow lead to fractures and faults in the sediments, and change the local permeability; in this respect mud diapirs are similar to salt diapirs. At the same time there is a significant difference between salt and mud diapirs: buoyancy, which is the main motive force of salt diapirism, also has a significant influence on the motion of a mud diapir. However, hydrocarbon gases, produced by deep burial of the mud diapir sediments, create an additional overpressure which can drive large volumes of mud vertically upward through the sedimentary section, thereby providing one of the main reasons for the spectacular eruptions over mud diapirs. There is a very

171

complicated mutual interdependence of diapiric processes caused by buoyancy and hydrocarbon generation (especially methane).

Driving mechanisms of mud diapirs are complex and various, as presumably reflected in the diversity of surface expressions. Excepting mud diapirs which are driven dominantly by forces of buoyancy, other forms of mud intrusions, named diatremes (Brown, 1990), are formed as a result of fluidization of parcels of unconsolidated rock. Quantitative evaluation of similar processes has been developed elsewhere (Ivanov and Guliev, 1986a, 1986b; Lerche et. al., 1996). The objective of this chapter is not to search for the driving forces of mud intrusions as diapirs or diatremes, but rather to use the results of quantitative basin modeling to study the impact of developing mud diapirs on the dynamical and thermal features of surrounding rocks, on hydrocarbon generation and accumulation under such conditions, and on the associated gas hazards from such mud diapirs. The South Caspian Basin has been chosen as a region of investigation, a classic region of mud diapirism and mud volcanism development. Application of the procedures is focused on the Abikh mud diapir, which is located on the only 12-second deep seismic profile available.

Using self-consistent quantitative modeling methods for diapirs and sediments (Lerche and Petersen, 1995; Lerche et al., 1996), the history of development of the Abikh diapir has been evaluated elsewhere (Bagirov and Lerche, 1997), based on its present-day shape, and on the depths and ages of sedimentary surfaces which were assessed from the seismic section. It was shown that the development of the diapir started at Middle Pliocene time, when very rapid sedimentation processes started. Assuming approximately horizontal deposition of sediments, the speed of rise of the diapir was evaluated. Then, using physical properties of mud and sediments, temperature anomalies and the stress and strain of the surrounding rocks were estimated. The question considered here is: What is the influence of these anomalies on hydrocarbon generation and accumulation? In the South Caspian Basin almost all structures (where oil and gas fields are located) are complicated by mud volcanoes or intrusions. In most structures oil accumulations are located on the flanks of mud bodies, while the bodies themselves are gas-charged. Questions that arise are: Is that phenomenon only random, or is it systematic caused in some way by the nature of a mud intrusion? Do the diapirs occur in places of high concentrations of natural gas, or are they excellent traps for gas? To investigate these problems a 2-D basin modeling technique was used. To circumvent questions about the nature and driving mechanisms of a diapir, data describing diapiric evolution were input to the codes. Fortunately, the program GEOPETII, which was used as a basic code, has a "KEEP" option, which allows one to stop the calculations at any time-step, and to then change any values of any parameters in the input data file. Thus one can insert the evolution of a diapir (more precisely, one can replace the existing rocks with the physical properties of "sloppy" mud at each time-step) from the bottom position of the diapir to the top position of the diapir. The calculations are then continued until the next time-step when, once again, one has information about the position of the diapir. The width of the Abikh diapir is approximately 10 km, and the form is almost pillar-shaped. Account was not taken of the precise shape of the present-day diapiric body, which is quite complex (with a small overhang on the crest and a large bulge in the left flank). The influence of these fluctuations is negligible when compared with the coarse spatial resolution used in the code. To illuminate the influence of a mud diapir the model was also run with precisely the same input parameters, but for structures undisturbed by a diapir. The relative importance procedure (Lerche et al., 1996) is used here to show the relative influence of the diapir on the different modeled outputs.

A. Geology and Input Parameters

The South Caspian Basin (SCB) contains sediments from Jurassic to Quaternary in age. Discovered oil and gas deposits are sited in reservoir traps ranging in age from Upper Cretaceous to Quaternary. The SCB has some peculiarities which, when taken together, make this basin unique from many viewpoints, including oil and gas content:

    1) The SCB has one of the thickest sedimentary covers known for any basin, up to 25-30 km according to seismic, magnetic and gravity data, making the SCB truly a unique sedimentary basin;

    2) The basement of the central offshore part of the SCB is a relict fragment of Tethyian oceanic crust, with basinal subsidence to 25-30 km being a result of collision connected with closure of the Tethys Ocean.

    3) In the central part of the SCB, extremely low values of geothermal gradients are found (1.0 to 1.8°C/100 m), compared to the more "usual" thermal gradients of 2.5-3.5°C/100 m in other basins, which puts the present-day depth limit for the "oil window" onset to about 9 km and the "gas window" onset to about 14 km;

    4) High excess fluid pressure is encountered pervasively throughout the SCB, with excess pressure evolution tied to the widespread development of mud diapirism and mud volcanism.

    As a basis for 2-D quantitative analysis of basin subsidence and sedimentation rates, a portion of the 12-second cross-section was used (Gambarov et al., 1993). (The results of the 2-D modeling along all of the section can be found elsewhere (Lerche et al., 1996)). Eight seismic-stratigraphic units (SSU), mentioned in Chapter 1, are picked out on this regional cross-section. A part of the 12-second seismic cross-section is shown on figure 7.1. Beginning from middle Pliocene time the sedimentation rate underwent a rapid increase of approximately one order of magnitude and reached turbidite (avalanche) values. Sedimentation rates for the middle Pliocene reached 1.3 km/My in the central abyssal part of the SCB (Lerche et al., 1996), which enhanced the development of huge mud structures, such as the Abikh diapir.

    Geological data taken from the 12-second seismic cross-section were used, as in chapter 3 of Lerche et al. (1996). The length of the cross-section provides a lateral distance of 60 km. Nineteen pseudo-wells, separated at 3.3 km spacing, were chosen along the seismic cross-section. With the aim to minimize edge effects, 10 km of shale section were added to both lateral section

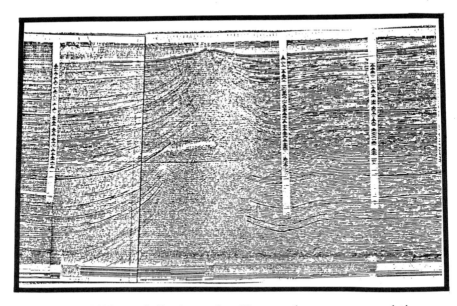

Figure 7.1. Abikh mud diapir on the 12-second two-way travel-time cross-section.

edges. Thus the total length of the section modeled reached 80 km. An extra 6 pseudo-wells were then added to accommodate for edge conditions, for a total of 25 pseudo-wells. The surface of seismic-stratigraphic unit SSU-2, which stretches relatively smoothly across the full cross-section distance of the Mesozoic sedimentary complex of the SCB, was taken as the basement surface. This surface is clearly observed around the Abikh diapir.

Seven of the SSU boundaries of Chapter 1 were chosen to be the present-day depths of the main stratigraphic surfaces with time, as follows:

1) the basement; 2) the Mesozoic surface; 3) the Eocene surface; 4) the Maikop (Oligocene - Lower Miocene) surface; 5) the Middle Pliocene surface; 6) the Upper Pliocene surface; 7) the Quaternary surface (sea bottom). All of these horizons exist across the total cross- section, which is 100% offshore.

The present-day depth values for both the basement and for the six sediment layer surfaces were determined in the 25 pseudo-wells along 80 km of the 12-second seismic cross-section. Then these SSU layers were divided into finer thickness strata in accordance with knowledge about stratigraphic and lithologic units from outcrop and drilling information in parts of the SCB. This procedure allows one to pick out 26 layers in the section (Fig. 7.2).

Due to the ability of GEOPETII to input a mix of up to 7 lithologic types at each grid point, the model lithology was tied to: (i) the lithology of the same age layers from outcrops and SCB wells; (ii) to breccia fragments

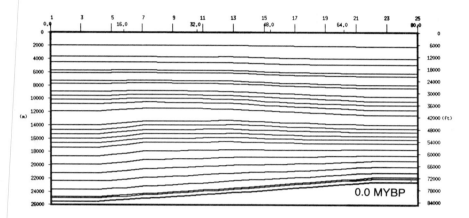

Figure 7.2. 2-D section layers indicating strata depths around the location of Abikh diapir with the diapir omitted.

from mud volcanoes; and (iii) in accord with the lithologic changes observed in the abyssal direction.

The lithological content of each stratigraphic unit can be found in Lerche et al., (1996, Chapter 3). The lithological content of mud, which constitutes the body of the diapir, was defined as shaly, with the same surface porosity, permeability, thermal conductivity and density as shale sediments. But the porosity is taken to be almost unchanged with depth, which allows one to inject a "sloppy" mud, with all the features of shale sediments, in places where the diapir is located and to construct a pseudo-dynamic model of the developing diapir.

Heat flows for the offshore SCB (except for the Apsheron-Balkhan zone) range between 0.5 - 1.0 HFU (1 HFU $\cong$ 42mW/m^2). The lack of data that might allow a better judgment about heat flow variation along the 2-D section forced an acceptance of an average present-day heat flow value of 0.65 HFU for all pseudo-wells along the 12-second deep profile. Based on tectonic evolution of the basin and on depositional thickness reduction in the past, it was assumed that the paleo-heat flow grows linearly with past geological time and, at the Jurassic time for the start of modeling, is taken as 1.65 HFU. The amount and type of kerogen for each layer, as well as heat flow and lithology content, are tabulated in Table 7.1.

Other factors used in the model are:

(i) The sea-bottom temperature for all deposits is assumed to be 9°C;

(ii) The fracture coefficient (the ratio of total fluid pressure to lithologic load at which rocks fracture) was taken to be 0.83;

Table 7.1. Input Information for Basin Modeling.

LAYER Index	Symbol	Age, MYBP	HFU	TOC %	Kerogen Type % I	II	Lithology (%)
26	QUAT	1.3	0.75	1.5	0	100	Sh=70, Sd=30
25	APSH	1.6	0.75	1.5	0	100	Sh=80, Sd=10, Lm=10
24	AKCH	1.8	0.75	1.5	0	100	Sh=90, Sd=10
23	MPL5	2.5	0.76	1.5	0	100	Sh=90, Sd=10
22	MPL4	2.7	0.76	0.3	0	100	Sh=10, Sd=90
21	MPL3	3.6	0.77	1.5	0	100	Sh=90, Sd=10
20	MPL2	3.9	0.78	0.3	0	100	Sh=10, Sd=90
19	MPL1	5.0	0.79	1.5	0	100	Sh=90, Sd=10
18	PONT	6.2	0.82	1.5	10	90	Sh=90, Sd=10
17	MIO3	10	0.85	1.5	50	50	Sh=80, Sd=20
16	MIO2	16	0.86	1.5	30	70	Sh=40, Sd=50
15	MIO1	23	0.90	3.0	20	80	Sh=90, Sd=10
14	OLIG	35	0.95	4.5	20	80	Sh=100
13	EOC3	39	0.98	0.7	0	100	Sh=70, Sd=20, Lm=10
12	EOC2	50	1.03	0.7	0	100	Sh=80, Sd=10, Lm=10
11	EOC1	57	1.05	0.7	0	100	Sh=50, Sd=20, Lm=20,Do=10
10	PLC2	61	1.07	0.7	0	100	Sh=50, Sd=20, Lm=20,Do=10
9	PLC1	65	1.10	0.3	0	100	Sh=40, Sd=20, Lm=20,Do=20
8	CR2C	74	1.15	0.7	0	100	Sh=60, Sd=10, Lm=20,Do=10
7	CR2T	97	1.20	0.7	0	100	Sh=60, Sd=10, Lm=20,Do=10
6	CR1V	124	1.30	0.7	0	100	Sh=60, Sd=10, Lm=20,Do=10
5	CR1L	145	1.40	0.7	0	100	Sh=60, Sd=10, Lm=20,Do=10
4	MALM	157	1.50	0.7	0	100	Sh=60, Sd=10, Lm=20,Do=10
3	DOGR	173	1.60	3.0	0	100	Sh=60, Sd=10, Lm=20,Do=10
2	VLUP	178	1.63	0.0	0	100	Cr=100
1	BSMT	180	1.65	0.0	0	100	Cr=100

Legend: Sh = Shale, Sd = Sand, Lm = Limestone, Do = Dolomite, Cr = Crystalline.

(iii) Because the model must be run for the recent high (avalanche) sedimentation rate stage with small time-steps (0.1 MY), and for the earlier stage of the section with a larger time-step (0.2 MY), an option in GEOPETII was used, called KEEP, which permits the code to be stopped, any or all values to be changed, and then the code restarted;

(iv) For the oil and gas generations parts of the model, the Tissot kinetic model was used which assumes six channels for kerogen degradation into oil and one channel for oil degradation into gas.

The base of the diapir was taken to be rooted in Cretaceous deposits, and diapiric growth with time was defined as in the previous chapter. As evaluated to the end of Maikopian time, there was no evidence of diapir evolution and the sedimentary layers were horizontal and non-deformed. The starting point of diapiric development was sometime in the Miocene or early Pliocene, and, by the beginning of Productive Series time (Middle Pliocene), a diapir existed of approximately 8-9 km height. At this time the top of the mud diapir was at a sub-mudline depth of 2-3 km. The further growth of the diapir was accompanied by huge rates of sedimentation; by the end of Middle Pliocene and Late Pliocene, in spite of high rates of diapiric uprise (to the beginning of the Quaternary, the diapiric pillar height reached 16-18 km), the top of the intrusion remained at a sub-mudline depth of 2-3 km, and, at present-day, is almost at the sediment surface (sea bottom). By the beginning of Middle Pliocene, the diapiric intrusion penetrated all Cretaceous, Paleogene and Lower Miocene (Maikopian) sediments. By the end of Pliocene time, the top of the diapir was in the upper part of the very thick Productive Series and, at present-day, penetrates all the sedimentary cover from Cretaceous to Quaternary. Figures 7.3 and 7.4 (a-d) show the rise of the diapir, consisting of high porosity material .

B. Model Results

To show the impact of the growing diapir on other processes in the sedimentary formations, the GEOPETII 2-D code was run twice, once for the case when formations are not disturbed by a diapir, and once for the case with the diapir. The Relative Importance procedure was modified to show the relative influence of one variable parameter on a set of outputs. Mathematical considerations of the modifications are given in Appendix A.

Figures 7.5a-d show how different output parameters react to inserting the body of the diapir at different time-steps. The Relative Sensitivity, defined in terms of relative influence on different output parameters, does not change significantly with time. The only really sensitive output parameter is Lopatin's TTI, which varies from very insensitive at the diapiric start time to extremely high sensitivity at present-day because of its strong dependence on temperature changes. The diapiric activity is particularly dominant in its influence on the fluids (and especially gas) flow rates, generation, accumulation and saturation; a lesser impact is felt by vitrinite reflectance, pressure, physical properties of the sediments (porosity, permeability, thermal conductivity), and by temperature. These remarks are appropriate for global relative sensitivity, which measures the influence on averages of parameters across the section at a given time-step. More interesting is to see how far the influence extends away from the location of the diapir. The uncertainty of different parameters across the section (as measured by the value of logarithmic variance, μ, see Appendix

178

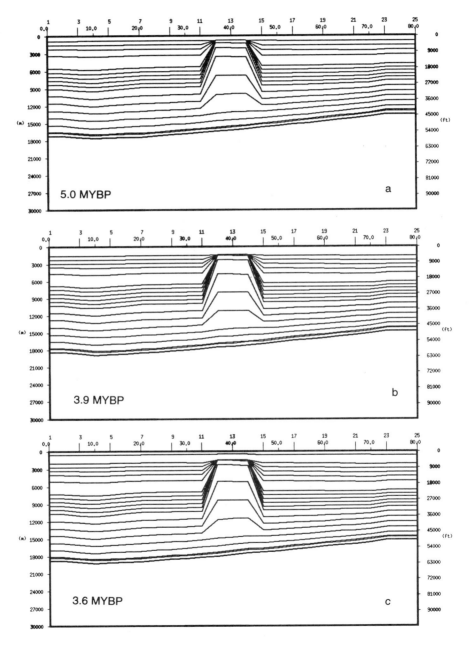

Figure 7.3. Development of the Abikh diapir at: a) 5.00 MYBP; b) 3.9 MYBP; c) 3.6 MYBP.

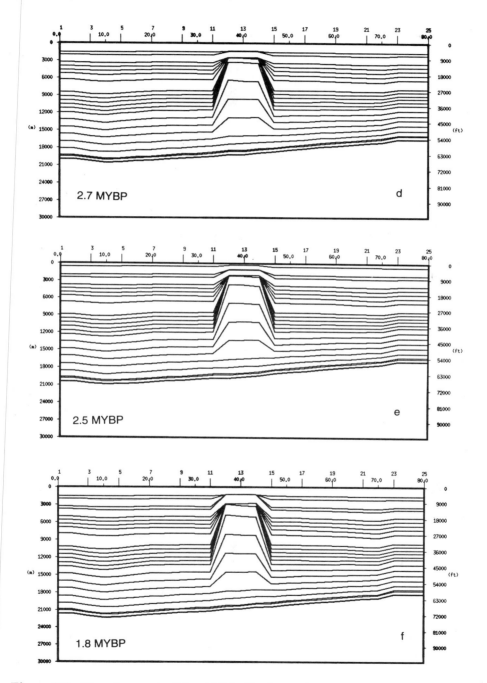

Figure 7.3. Development of the Abikh diapir at: d) 2.7 MYBP; e) 2.5 MYBP; f) 1.8 MYBP.

180

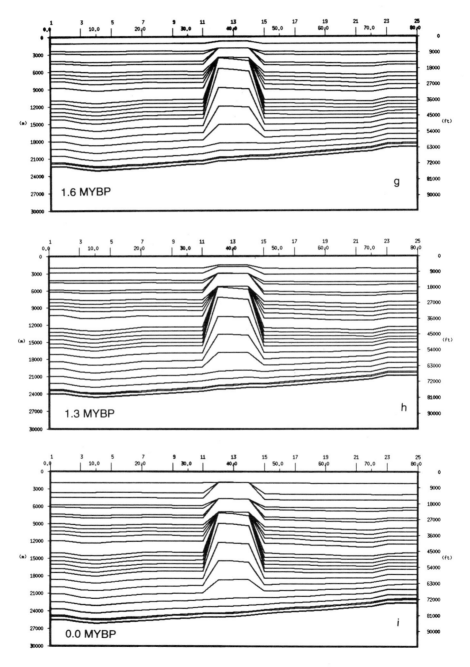

Figure 7.3. Development of the Abikh diapir at: g) 1.6 MYBP; h) 1.3 MYBP;
i) present-day.

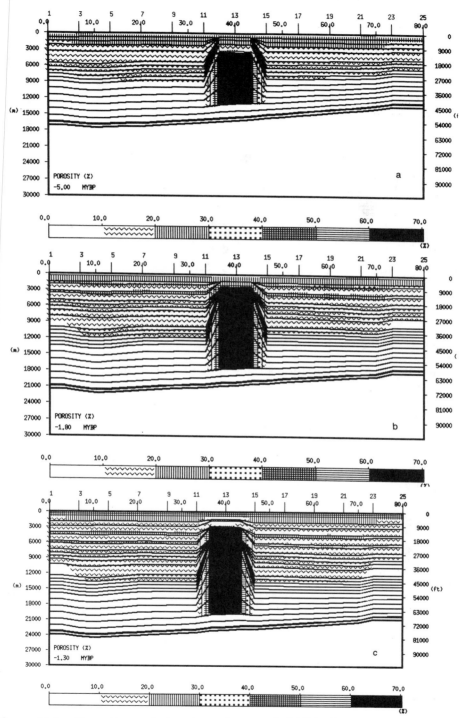

Figure 7.4. Porosity variation in and around Abikh diapir across the 2-D section: a) at the beginning of the Middle Pliocene; b) at the end of the Middle Pliocene; c) at the beginning of the Quaternary.

182

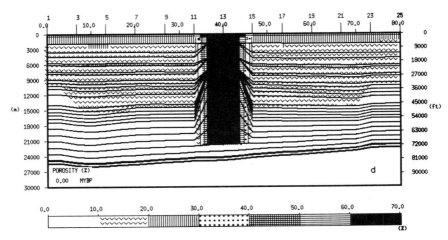

Figure 7.4. Porosity variation in and around Abikh diapir across the 2-D section: d) at present-day.

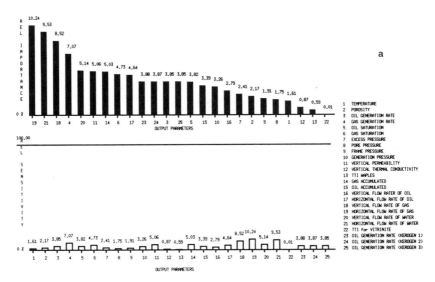

Figure 7.5. Relative influence of the mud diapir on the output parameters of 2-D basin modeling at: a) the beginning of the Middle Pliocene.

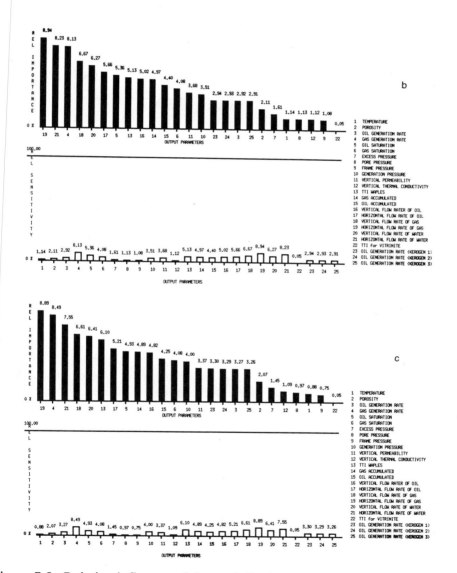

Figure 7.5. Relative influence of the mud diapir on the output parameters of 2-D basin modeling at: b) the end of the Middle Pliocene; c) beginning of the Quaternary.

184

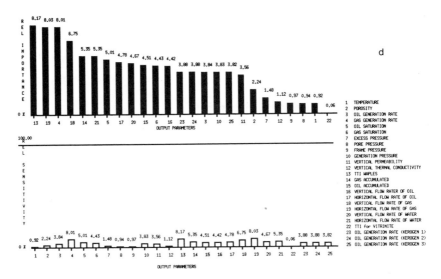

Figure 7.5. Relative influence of the mud diapir on the output parameters of 2-
D basin modeling at: d) present-day.

A) shows the influence of the diapir on the sedimentary formations with depth
and laterally. The pictures of influence are surprisingly different.

Because the body of the diapir was inserted by replacement at the
location of the mud intrusion, one can expect that parameters such as porosity,
permeability, and thermal conductivity will be different mainly in the zone of
mud intrusion; values of μ greater than zero occur only at grid-points inside the
diapiric body for these parameters (see for instance, Fig. 7.6). The temperature
has a more complex interrelation with conductivity of the sediments, which is
why the uncertainty picture of the temperature produces concentric circles
around the diapir (Fig. 7.7), but the values of uncertainty are low at about a
radius away from the diapir. The higher values of uncertainty correspond to
more sensitive parameters, like Lopatin's TTI and the generation of
hydrocarbons (Figs. 7.8 and 7.9). In spite of high uncertainty values (up to μ
≈ 2) for gas generation rates, the picture of uncertainty remains concentric
circles as is also the case for oil generation rates (but the uncertainty is then
clearly seen only for the upper part of the section). Parameters such as fluid
flow rates and hydrocarbon accumulations, being highly related with lithology,

have high values of uncertainty almost everywhere across the section (Figs. 7.10, 7.11). To show how far this influence extends, the scale and length of the cross-section were changed. Figures 7.12, 7.13 and 7.14 show the values of μ for the same parameters across a 160 km length cross-section. The location of the mud intrusion corresponds to pseudo-well location 9, where one can see that the influence of the mud diapir extends to more than 60 km from the diapir, i.e. about 6 diameters of the diapir.

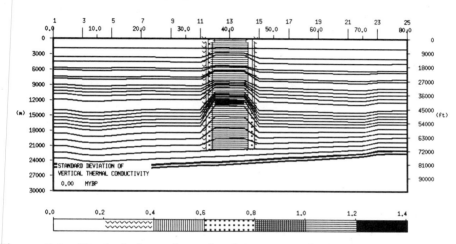

Figure 7.6. Vertical thermal conductivity uncertainty across the section at present-day as measured by the logarithmic slope factor, μ.

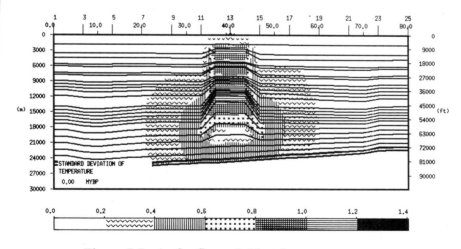

Figure 7.7. As for figure 7.6 but for temperature.

Now consider values of specific modeled parameters under the influence of diapiric activity.

Two very important parameters, which influence the whole basinal development process, are excess pressure and temperature. Figure 7.15 shows

186

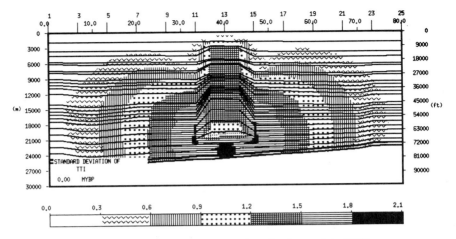

Figure 7.8. As for figure 7.6 but for Lopatin's TTI.

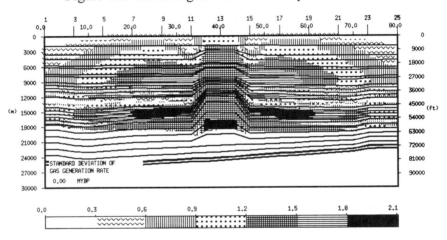

Figure 7.9. As for figure 7.6 but for gas generation rate.

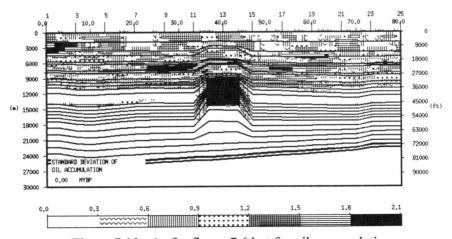

Figure 7.10. As for figure 7.6 but for oil accumulation.

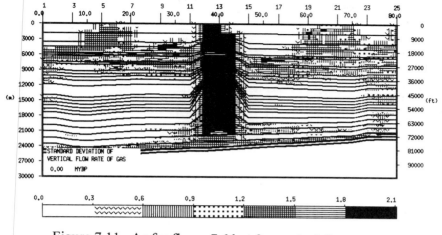

Figure 7.11. As for figure 7.6 but for vertical flow rate of gas.

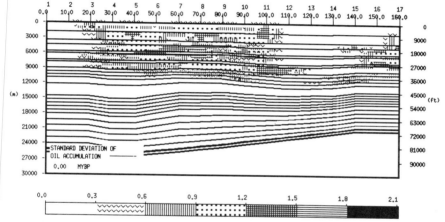

Figure 7.12. Oil accumulation uncertainty across 160 km length cross-section at present-day as measured by the logarithmic slope factor, μ.

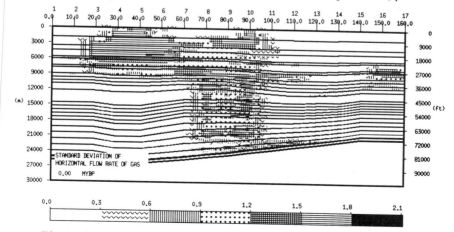

Figure 7.13. As for figure 7.12 but for horizontal flow rate of gas.

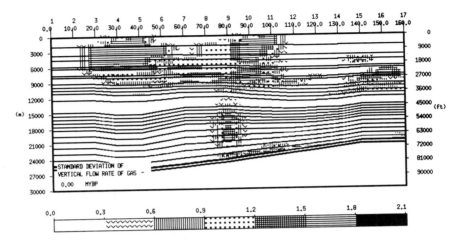

Figure 7.14. As for figure 7.12 but for vertical flow rate of gas.

excess pressure in the sediments both in the absence of the diapir and with the diapir present. The reason for the negligible excess pressure inside the body of the diapir is due to the lack of compaction. The temperature anomalies created by the mud diapir have been considered elsewhere (Lerche et al., 1996). The low conductivity of the body of the diapir forces the heat flow to avoid the structure (Figure 7.16), as a result the top of the diapir is much cooler than surrounding sediments. Figures 7.16(a-d) show the patterns of the temperature anomalies at different time-steps, corresponding to different stages of development of the diapir.

The excess pressure and temperature anomalies mainly control the anomalies in oil and gas generation, migration and accumulation. Oil generation rates were high during the last two million years, and reached peak values at about the beginning of the Quaternary period (Bagirov et al., 1997). The temperature anomalies contribute to the oil generation being higher around the diapir, while there is only very low oil generation inside the body of the low conductivity diapir. This phenomenon is clearly observed in figures 7.17, 7.18, and 7.19. Approximately the same phenomenon is observed in the gas generation rates (Figs. 7.20, 7.21, 7.22, 7.23), with the major difference being that there is some generation of gas, especially at the early stages of development, in the body of the diapir.

The difference in excess pressure created by the growing mud intrusion leads to very interesting flow pictures as follows.

1. Water Flow

The range of the projection of the water flow rates in the vertical direction at the beginning of the Middle Pliocene (when the diapir just started to rise) varies from 530 m/Ma to 1000 m/Ma in the upward direction. And these fast streams

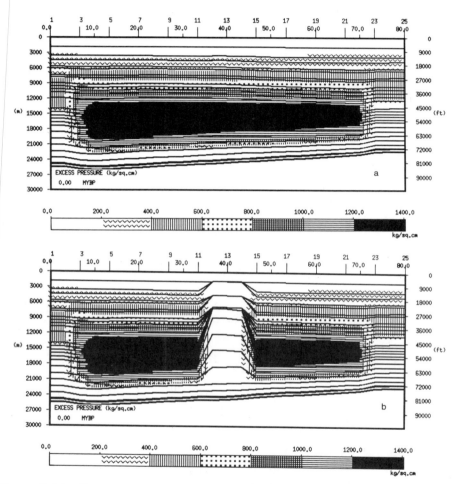

Figure 7.15. Excess pressure across the 2-D section at present-day: a) for undisturbed sediments; b) for sediments with the diapir included.

of flow occur in the sandy lithologies, which constitute the main fraction of the Miocene. However, as well as the flow of water caused by the process of compaction, there is also the migration of water into the body of the diapir, with a speed of 100-700 m/Ma, from the top part of the diapir (Fig. 7.24a). The horizontal flow rates at that time are steadier, and vary in the range up to 30 m/Ma in both directions (Fig. 7.24b). High values of water flow rates in the horizontal direction occur around the border with the diapir, where values of

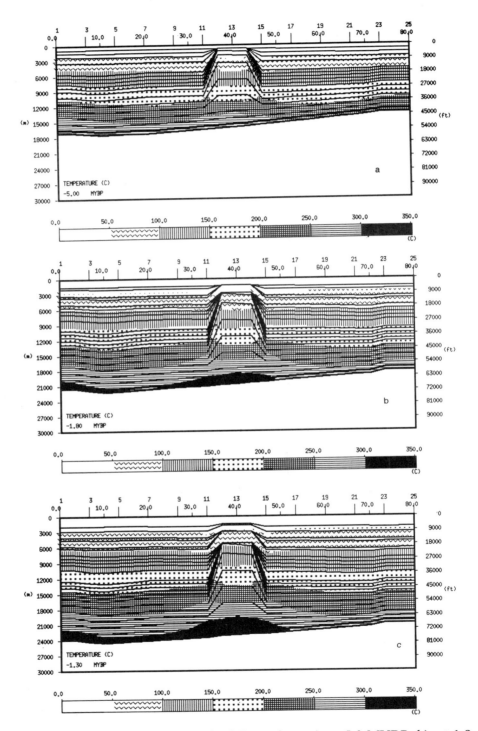

Figure 7.16. Temperature across the 2-D section: a) at 5.0 MYBP; b) at 1.8 MYBP; c) at 1.3 MYBP.

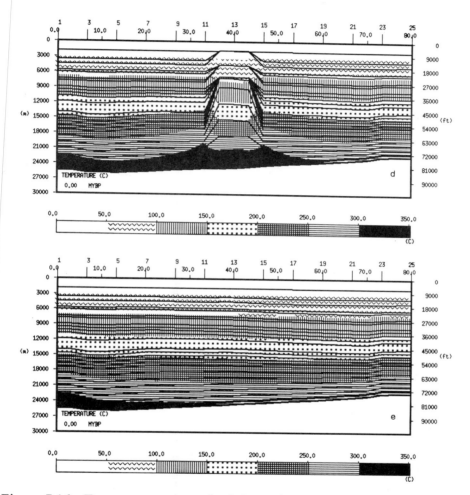

Figure 7.16. Temperature across the 2-D section: d) for sediments with the diapir included at present-day; e) for undisturbed sediments at present-day.

100-200 m/Ma are reached. The same patterns of water flow occur at the end of the Middle Pliocene (Fig. 7.25a,b), but the ranges of the flows are higher because of sandy layers in the Middle Pliocene deposits, which were then still very shallow; the vertical and horizontal flow rates reach almost 700,000 m/Ma in these layers (70 cm/year). The dominant flow to the diapiric body is from these sandy layers. Because of these high speeds of water streams it is very difficult to trace the flow rates in the shaly part of the section at the same time.

192

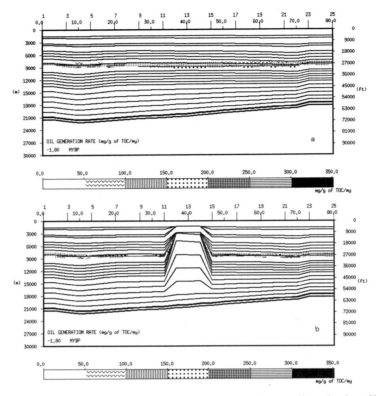

Figure 7.17. Oil generation rate at 1.8 MYBP: a) for undisturbed sediments; b) for sediments with the diapir included.

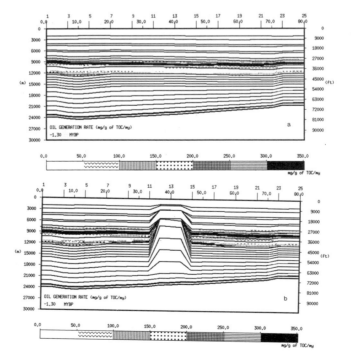

Figure 7.18. As for figure 7.17 but at 1.3 MYBP.

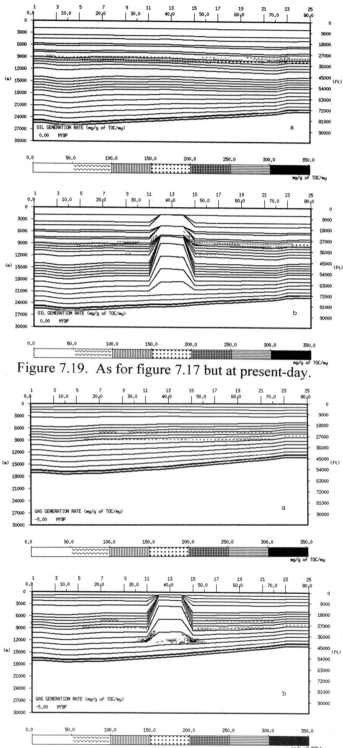

Figure 7.19. As for figure 7.17 but at present-day.

Figure 7.20. Gas generation rate at 5.0 MYBP: a) for undisturbed sediments; b) for sediments with the diapir included.

194

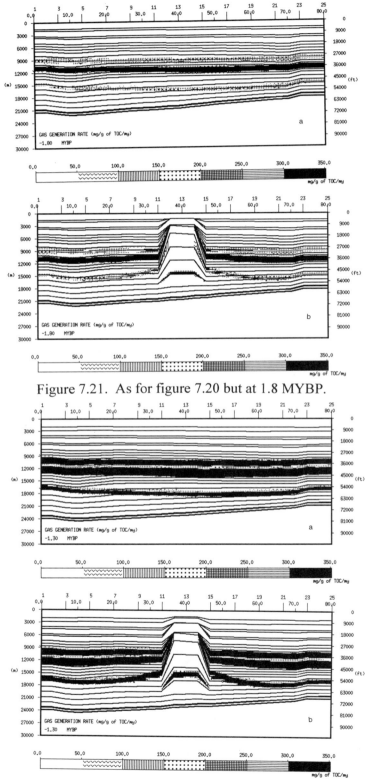

Figure 7.21. As for figure 7.20 but at 1.8 MYBP.

Figure 7.22. As for figure 7.20 but at 1.3 MYBP.

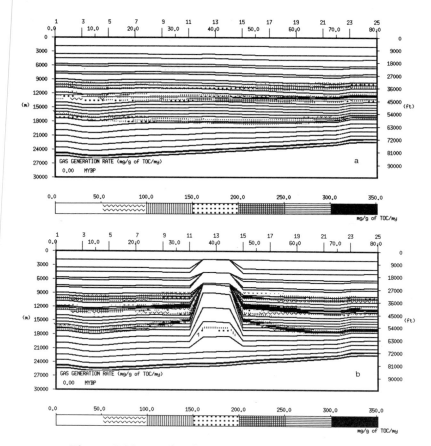

Figure 7.23. As for figure 7.20 but at present-day.

At the beginning of the Quaternary, horizontal water flows inside the body of the diapir are observed (Fig. 7.26). And, at present-day (Fig. 7.27), lateral flow of water takes place with rates up to 60 cm/year from sandstones of the Productive Series (Middle Pliocene). Comparing the flow patterns of oil and water, one can observe a close similarity between them (Figs. 7.28, 7.29, 7.30, 7.31). Despite the higher viscosity of oil compared to water, the flow rates of oil in the upper part of the section (where oil exists) are higher than the water rates. And, as in the water case, if the major direction of flow was vertical in the early stages of diapiric activity, then at the later stages the direction of flow towards the diapiric structure is more dominant.

2. Gas Flow

Gas flow is essentially different from those of water and oil due to the differences in surface tension and low gas viscosity. At the first stages of diapiric development (Fig. 7.32a) there was a rapid uprising of gas inside the diapir, with the rates of vertical flow of gas reaching 30-40 cm/yr. At the same time, there was also some lateral gas flow (Fig. 7.30b) of up to 300-400 m/Ma. Starting from the end of Middle Pliocene through to present-day, the lateral

196

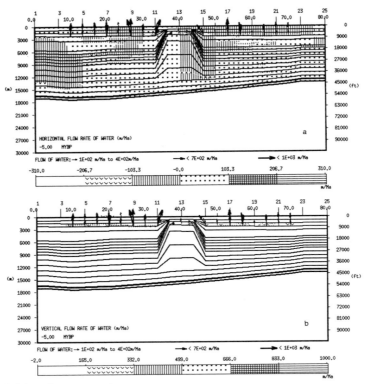

Figure 7.24. Flow rate of water at 5.00 MYBP: a) horizontal; b) vertical.

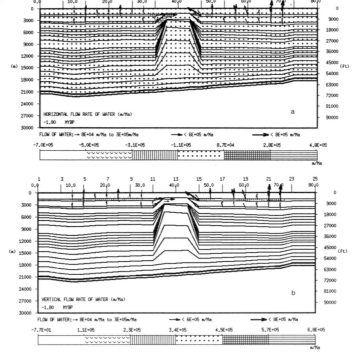

Figure 7.25. As for figure 7.24 but at 1.8 MYBP.

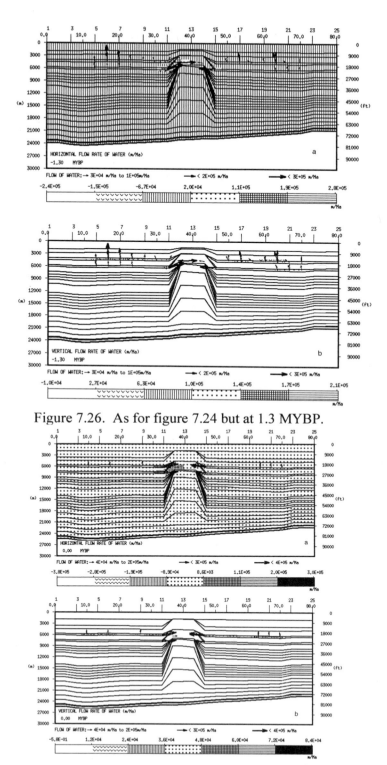

Figure 7.26. As for figure 7.24 but at 1.3 MYBP.

Figure 7.27. As for figure 7.24 but at present-day.

198

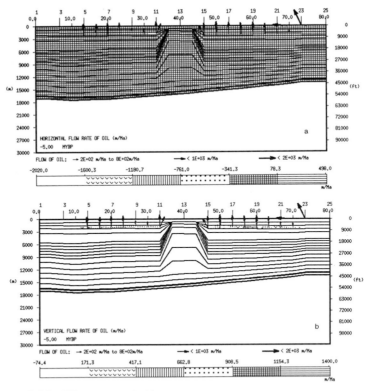

Figure 7.28. Flow rate of oil at 5.00 MYBP: a) horizontal; b) vertical.

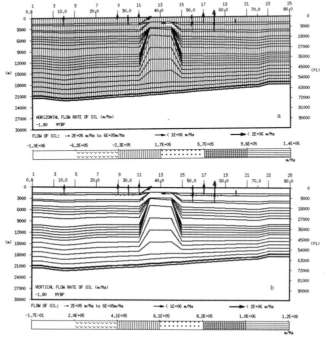

Figure 7.29. As for figure 7.28 but at 1.8 MYBP.

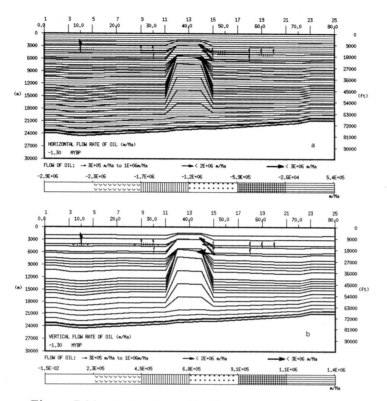

Figure 7.30. As for figure 7.28 but at 1.3 MYBP.

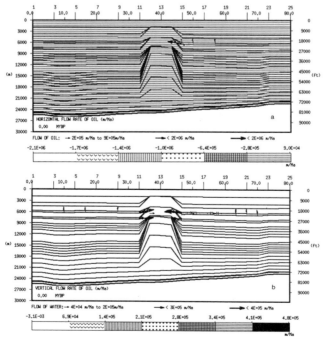

Figure 7.31. As for figure 7.28 but at present-day.

200

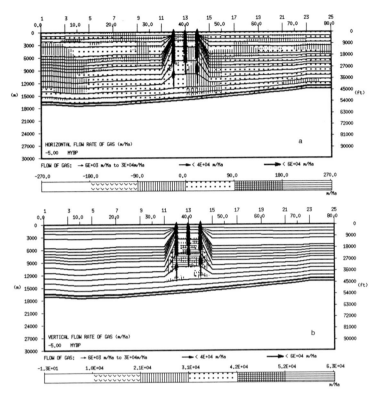

Figure 7.32. Flow rate of gas at 5.0 MYBP: a) horizontal; b) vertical.

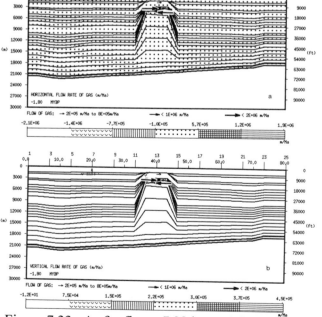

Figure 7.33. As for figure 7.32 but at 1.8 MYBP.

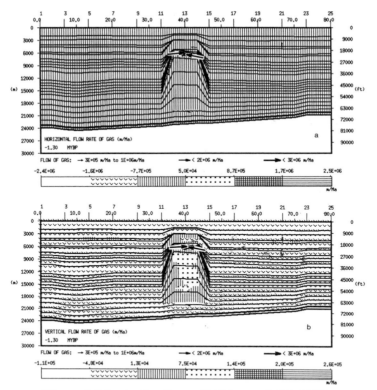

Figure 7.34. As for figure 7.32 but at 1.3 MYBP.

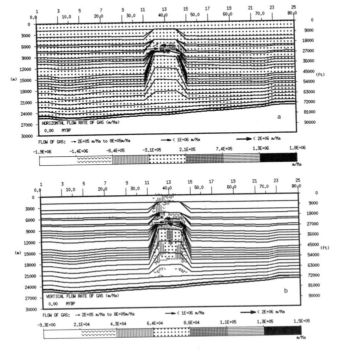

Figure 7.35. As for figure 7.32 but at present-day.

202

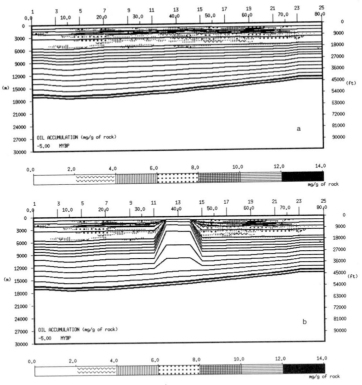

Figure 7.36. Oil accumulation across the 2-D section at 5.00 MYBP: a) undisturbed sediments; b) sediments with the diapir included.

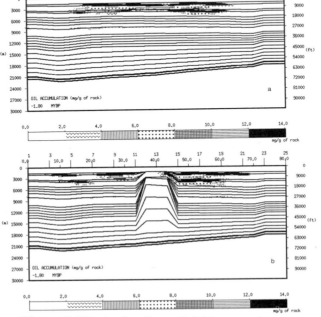

Figure 7.37. As for figure 7.36 but at 1.8 MYBP.

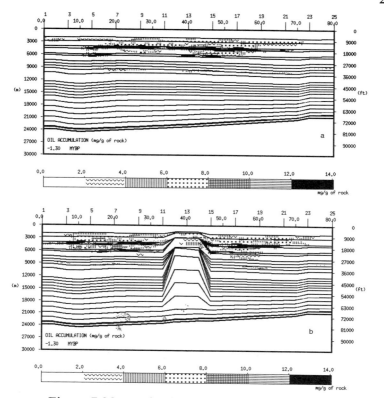

Figure 7.38. As for figure 7.36 but at 1.3 MYBP.

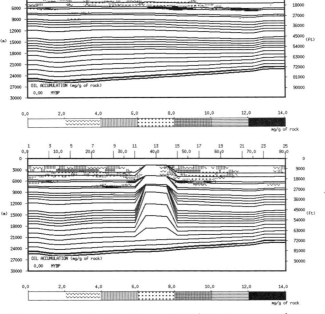

Figure 7.39. As for figure 7.36 but at present-day.

204

direction of gas flow has been dominant; inside the mud intrusion the vertical rate of gas flow was around 20-40 cm/year, at the same time the gas flow into the diapiric body was around 1m/year, which brings a considerable mass of gas into the diapir.

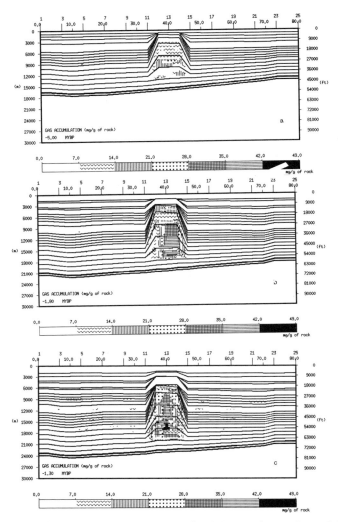

Figure 7.40. Free-phase gas content across the 2-D section: a) at 5.0 MYBP; b) at 1.8 MYBP; c) at 1.3 MYBP.

The anomalies in hydrocarbon generation, as well as migration of fluids, define the anomalies in oil and gas contents in the sediments. Compare the oil accumulation patterns with those under the assumption of sediments undisturbed by the diapir. From the first stage through to present-day there is only a small amount of oil inside the body of the diapir. On the flanks of the diapir, the free-phase oil content is slightly higher than it would be if no diapir was present.

The gas content is different. Figure 7.41a shows that in the undisturbed situation, the free-phase gas content in the sediments varies in the range 0-14 mg/g rock. However, in the presence of a diapir, the body of the diapir is filled by gas. The content of free-phase gas is much higher than in surrounding sediments (Fig. 7.41b) and reaches values of 50 mg/g rock (Fig. 7.41c). If the diapiric crest intersects an open fault, the pressure conditions in the diapir change and involve more gas, so the free-phase gas content is then higher - up to 51.5 mg/g rock, with values of 30-40 mg/g rock almost everywhere in the diapir (Fig. 7.41d). Such high values of the free-phase gas content have occurred from the beginning of diapiric activity (Fig. 7.40a,b,c). The same situation happens with the saturated gas content. Gas saturation is estimated to be below 20% if no diapir is taken into consideration (Fig. 7.45a). However, the presence of a diapir, and especially in combination with an open fault, increases the saturation value to around 30% and even higher in some zones inside the mud diapir intrusion (Fig. 7.45b,c). Figures 7.42, 7.43 and 7.44 show that such a phenomenon is typical for all time-steps.

Effectively what is happening physically is that the mud diapir retains porosity, and so a permeability closer to that of shales at deposition due to the early gas-charging that provided buoyancy. The surrounding shales compact,

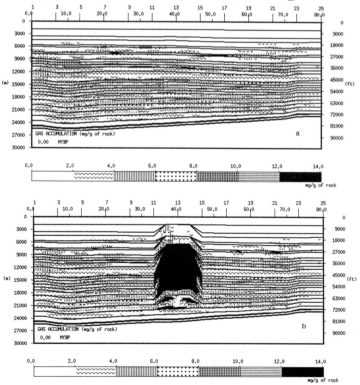

Figure 7.41. Free-phase gas content across the 2-D section at present-day: a) undisturbed layers case; b) with mud diapir.

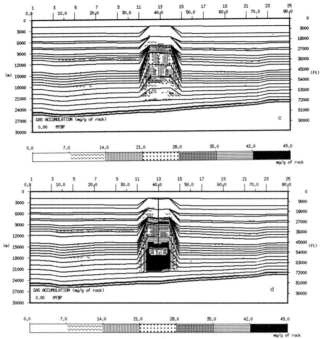

Figure 7.41. Free-phase gas content across the 2-D section at present-day: c) with mud diapir, but on a scale ranging up to 50 mg/g rock; d) with mud diapir and open fault.

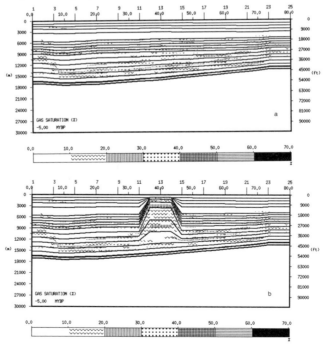

Figure 7.42. Gas saturation across the 2-D section at 5.0 MYBP: a) undisturbed layers case; b) with mud diapir.

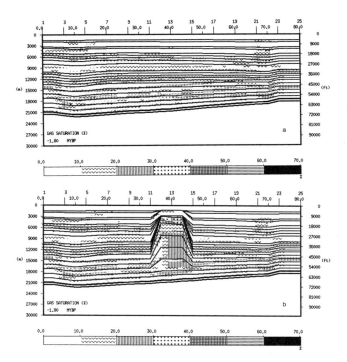

Figure 7.43. As for figure 7.42 but at 1.8 MYBP.

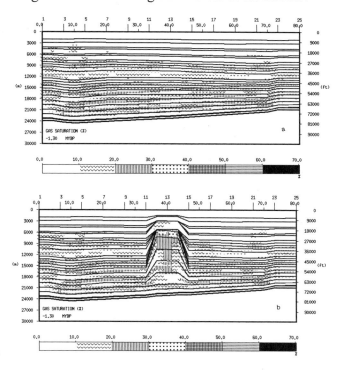

Figure 7.44. As for figure 7.42 but at 1.3 MYBP.

208

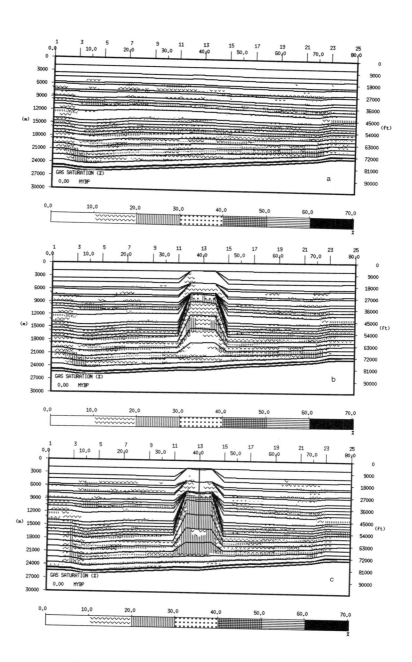

Figure 7.45. Gas saturation across the 2D-section at present-day: a) undisturbed layers case; b) with mud diapir; c) with mud diapir and open fault.

thereby lowering both their porosities and permeabilities. Any hydrocarbons generated in the shales now have a primary migration capability, driven by capillary pressure, to the more permeable mud diapir. Accordingly gas-charging of the diapir increases and the cycle repeats. Oil also migrates to the diapir, but at a slower rate than gas. In addition, the relative permeabilities and solubilities of gas and oil differ significantly, with the consequence that gas is easily placed in the diapir in water solution but oil, being almost incompressible and of low solubility, has difficulty displacing the gas-charged diapiric waters. Hence oil tends to accumulate more on the flanks of the diapir rather than in the core of the diapir.

In addition, the compaction of shales surrounding a diapir, and the late massive deposition in and after the Miocene, raise the excess fluid pressure in the shales to extremely high values (Tagiyev et al., 1996). However, in the diapir (with its underlying buoyancy pressure support) there is no need for such high fluid pressure. Thus a lateral excess pressure gradient is created, which also drives fluids (water, oil and gas) towards the higher porosity, greater permeability diapir region.

The release of pressure in a mud diapir by an overlying open fault, connected to the crest of the diapir, has several important consequences. First, the solubility of gas in water decreases as the pressure is released, causing exsolving of gas to free-phase gas. Because of volume conservation, the gas/water/mud mix then spurts out of the open conduit (fault), much as a froth of champagne and carbon dioxide is emitted from a bottle upon pressure release by cork removal. Thus a greater free-phase gas content occurs in the diapir, but the diapir is then also more subject to volcanic activity and mud flow production.

Second, the release of gas from solution causes a cooling of the mud (Joule-Thomsen effect) which changes the pressure-solubility behavior of the diapiric crestal region, and also lowers the near-crestal thermal conductivity of the diapir, causing a greater cooling of the crest. Indeed, some of the gas can then go back into solution to be available for further "champagne frothing" as other conduits for pressure release become available.

At the same time, gas is continually being driven into the diapir and gas-charged mud deep in the diapiric stem is still rising through the sediments. Thus the changes in dynamic evolution of a diapir, and the intermittent volcanic activity, depend on a complex interlocking of constantly shifting behaviors and contributions. Thus, it would seem difficult to predict precisely the occurrences of volcanic activity except in a statistical manner.

C. Hazard Aspects

The diapirs are just one of the forms of so called "mud intrusions". Laterally extended zones of decompaction are mentioned by Guliyev (1996) and are indicated by a zone of low velocity seismic waves, and are probably another example of massive "mud bodies". Probably there are also small structures or lenses, which consist of unconsolidated shale. It is difficult to say what the initial reason is for the formation of such structures. They represent special features of specific shales, or transformation of some clay minerals to others (for instance, montmorillonite to kaolinite) associated with dehydration of the minerals (The observations of waters of mud volcanoes do show very low mineralization). Gas generation can also play a role in this regard. But it is clear from the data-tied models that once such zones are formed they become traps for gas, which can fill the trap in free-phase, but does so mostly in solution. And, if a well encounters such zones while being drilled, then the pressure in the zone drops, bringing to the fore the "champagne frothing" effect, or microeruption. This effect can easily destroy a well. This process is one of the hazards which drillers face in the South Caspian Basin. And such damages are caused not only by technical incompetence, but also (most of the time) by geological conditions. So very fine-scale sonic logging, together with microseismic and seismic research of an area are necessary in order to detect such highly hazardous zones of low seismic velocity and high gas saturation.

D. Conclusions

The problems of mud diapirism activity and associated oil and gas presence has long been of interest to petroleum geologists because most of the mud volcanoes of the South Caspian Basin are associated with oil and gas fields. Mud volcano activity, both in its quiet stage as well as during eruptions, involves a high amount of methane gas, water and mud. And just a few volcanoes emit pure oil through their gryphons.

The questions which are always asked are: Do the mud volcanoes create oil and gas accumulations, or do they just destroy existing pools? What is the role of the moving mud in the formation of oil and gas fields? The model in this chapter provides answers to these questions. A diapir starts to rise under buoyancy, attracting fluids from surrounding sediments. Gas, being more mobile because of its low viscosity, first fills the volume of the mud intrusion. On the other hand, oil which flows with gas, as well as oil generated "in situ" in the diapir, transform to gas very quickly because of temperature anomalies created by the intrusion. Therefore, a mud diapir can be considered as a huge accumulation of gas, whereas the oil is concentrated on the flanks of the diapiric structure. The model also gives some ideas on the origin and driving forces of mud diapirs.

The starting point of the diapir begins with the formation of a zone of decompaction, perhaps as a consequence of a high rate of gas generation. In the South Caspian Basin this starting time coincided with the onset of very rapid subsidence of the basin (Lerche et al., 1996). This under-compacted zone started to rise under buoyancy.

Surprisingly, the excess pressure inside the diapiric intrusion is not high, and is less than in surrounding formations. As a result, fluids flow toward the body of the diapir, so that the interior of the diapir retains the features of sloppy mud with gas-charging of interstitial waters. The body of the diapir then grows, increasing the buoyancy forces, which further drive the diapir. Late stage, massive sedimentation in the last 3 million years implies a continued growth of the diapir (which grows about 25-30 km in about 15 Myr), but with the diapiric crest held at about 2-3 km below the sediment surface until very recently when sedimentation slowed. High speeds of sedimentation make the system more unstable. So this complex set of processes drives the diapir until it reaches the sediment surface (if the sedimentation rates are less than the upward speed of the diapir). Gas-charging of the diapir has continued until the present-day. Leakage of mud and gas through a crestal fault or fracture leads to explosive volcanic activity, with a re-building of gas pressure in the diapir thereafter. Oil, being less soluble in water than gas, tends to accumulate on the flanks of the diapir; and any oil which is trapped in the diapiric core is quickly converted to gas.

The thermal influence of the diapir extends laterally to about 2-4 diapir diameters into the surrounding sediments, while the diapir causes hydrocarbon flow anomalies and formation bed distortions out to about 6 diapir diameters.

The surface expression of the diapir depends on the tectonic complexity of the area. If there are open faults and fractures which cross the body of the diapir, then gas and mud can penetrate through the faults, and so form gryphons and salses on the surface. If the rates of gas release are not enough, then the pressure increases in the diapir and eruptions of mud volcanoes occur. For the Abikh mud diapir, with a root almost 30 km deep and a width of around 10 km, it would seem that the dynamical development portrayed here captures the essential elements of diapiric rise in association with gas accumulation.

Appendix A

Numerical application of cumulative probability procedures to basin analysis problems have been examined (Lerche et al., 1996) on three considerations: first is knowledge (or surmise) of the underlying, intrinsic, probability distributions for each parameter being used in the computation; second is knowledge (or surmise) of how multiple parameters combine to provide a probability distribution for each computational run, and how to extract the cumulative probability behavior of each required output from knowledge of the variations in each input variable parameter; third is the pragmatic concern of how to operate numerically to obtain the desired statistical information on attributes of interest.

In this Appendix we show how to use these techniques to evaluate the relative influence of the change of one input parameter on the set of output parameters.

A. Mathematical Considerations

In order to make full use of the techniques of probability measures, knowledge would be required of the intrinsic probability of obtaining an event. This knowledge is, by and large, based on data to do with scientific conditions. But the data available to work with are very limited at the early exploration assessment stage, usually imprecise or derived by analogy, or dependent on conditions about which little knowledge (as opposed to surmise) is available.

For these reasons, approximations and assumptions are introduced in attempts to obtain relatively robust estimates of data-related quantities from which some form of assessment can be made.

1. Single Parameter Distributions

By and large, parameters or variables that are not too well known (and we shall be more precise later) are usually treated as randomly varying in some manner around a mean value. The random component is customarily represented by a frequency distribution histogram which provides the relative number of times a parameter has been observed in a given interval range compared to all interval ranges. Customarily, the area under the histogram is normalized to unity so that the frequency distribution then provides an approximate empirical assessment of the probability of occurrence of a parameter in every interval range, based on the data to hand.

Often, interest centers not so much on the frequency distribution (e.g. what is the probability that the porosity is between 10% and 20%?) but rather on the *cumulative* frequency distribution (e.g. what is the probability that the porosity is greater (less) than 10%?).

In one case (greater than) the cumulative frequency distribution histogram is usually the residual fractional area still lying beyond a parameter value. In the other case (less than) the measure is the fractional area contained up to the parameter value. This is a mutually exclusive set (i.e. the probability of a porosity greater than 10% plus the probability of a porosity less than 10% must sum to unity). Thus if $p(x)dx$ measures the frequency distribution (normalized) of occurrence of x in the range x to $x + dx$, then:

$$P(y > x) = \int_{y}^{\infty} p(x)dx \qquad (A1)$$

measures the cumulative frequency distribution i.e. the chance of exceeding a particular value y; while:

$$P(y < x) = \int_{0}^{y} p(x)dx \equiv \int_{0}^{\infty} p(x)dx - \int_{y}^{\infty} p(x)dx \equiv 1 - P(y>x) \qquad (A2)$$

measures the chance of *not* exceeding a particular value y.

In a large number of circumstances in exploration assessments, it is often difficult to provide the frequency distribution even in a rough form. For that reason *moments* of the underlying distribution are often used as approximations. The mean value, $E_1(x)$, of x for a frequency distribution $p(x)dx$ is:

$$E_1(x) = \int_{-\infty}^{\infty} xp(x)dx \qquad (A3a)$$

while the mean square value $E_2(x)$ is:

$$E_2(x) = \int_{-\infty}^{\infty} x^2 p(x)dx \qquad (A3b)$$

The variance, σ^2, around the mean is given by:

$$\sigma^2 = E_2(x) - E_1(x)^2 \geq 0 \qquad (A3c)$$

where σ is the standard deviation.

In most of the situations we shall deal with, only multiple powers of distributions will be needed, defined by:

$$E_j(x) = \int_{-\infty}^{\infty} x^j \, p(x)dx \qquad (A4)$$

We will also deal with the median value, $x_{1/2}$, of a frequency distribution defined as that value of x such that:

$$P(y < x) = P(y > x) = 1/2. \qquad (A5)$$

and with the mode x_m (for a unimodal distribution) defined as that value of x at which $p(x)$ has its maximum value. The log-normal distribution plays fairly dominant roles in exploration strategy assessment., and occurs physically in many situations ranging from the areal size distribution of sunspots to lease sale bid distributions. The *normal* distribution cannot be appropriate when there is a constraint on a variable e.g. area cannot be negative, bid values must be positive. Under such conditions, empirical evidence suggests that an approximate measure of <u>cumulative</u> frequency distribution is provided by a log normal behavior with:

$$P(x|x_{1/2}, \mu) = \frac{1}{2} [1 + \mathrm{erf} \, (\ln(x/x_{1/2})/2^{1/2}\mu)] \qquad (A6)$$

with the mean value of x, $E_1(x)$, given through:

$$E_1(x) = x_{1/2} \exp (\mu^2/2), \qquad (A7a)$$

the mode value by:

$$x_m = x_{1/2} \exp (-\mu), \tag{A7b}$$

and the variance in x, $E_2(x) - E_1(x)^2 \equiv \sigma^2$, given by:

$$\sigma^2 = E_1(x)^2 [\exp(\mu^2) - 1], \tag{A7c}$$

where $x_{1/2}$ is the median value. At $x = x_\sigma \equiv x_{1/2} \exp (\mu)$ we have $P(x_\sigma|x_{1/2},\mu) = 0.84$, while on $x = x_m (\equiv x_{1/2} \exp(-\mu))$, we have $P(x_m|x_{1/2},\mu) = 0.16$. In this case note that $P = 1/2$ on $x_{1/2}$, but $P \cong 0.68$ on $x = E_1(x) > x_{1/2}$ (more exactly $E_1(x)$ is at a probability of about $(50+17\mu)\%$).

Empirically, it is often difficult, if not impossible, at the beginning of an exploration project to obtain enough information to determine the precise shape of the frequency distribution of a particular parameter or variable. Indeed, quite often it is considered a fairly good achievement to be able to estimate a likely minimum, x_{min}, a likely maximum, x_{max}, and a likely most probable value, x_p, for a parameter. A rough idea of relevant mean and variance can then be obtained from Simpson's triangular rule.

For the triangular distribution one has the estimates:

$$E_1(x) \cong \frac{1}{3} (x_{min} + x_p + x_{max}) \tag{A8a}$$

$$\sigma^2 \cong \frac{1}{2} E_1(x)^2 - \frac{1}{6} [x_{min} x_{max} + x_p (x_{min} + x_{max})] \tag{A8b}$$

$$E_2(x) = E_1(x)^2 + \sigma^2. \tag{A8c}$$

If it is further assumed that the variable is log normally distributed, it is possible to work equations (A7) in reverse to obtain estimates of μ, x, $x_{1/2}$ and x_m. Thus from equation (A7c) we obtain:

$$\mu = [\ln \{1 + \sigma^2/E_1(x)^2\}]^{1/2} \tag{A9a}$$

and then, from equations (A13a) and (A13b), we can estimate:

$$x_{1/2} = E_1(x)\exp(-\mu^2/2) \equiv E_1(x)[1 + \sigma^2/E_1(x)^2]^{-1/2} < E_1(x) \tag{A9b}$$

and the values:

$$x_\sigma = x_{1/2} \exp[(\ln(1 + \sigma^2/E_1(x)^2))^{1/2}] \tag{A9c}$$

$$x_m = x_{1/2} \exp[-(\ln (1 + \sigma^2/E_1(x)^2))^{1/2}] \tag{A9d}$$

2. Numerical Procedures

Empirically, the determination of μ^2, which controls the slope of the cumulative log probability curve at each coordinate, proceeds as follows.

Consider, first, that p is a variable component in a parameter vector. The output of a specific positive quantity, R, from the model is then available at spatial coordinates x, at time t, and is dependent on the specific values, p, used for the parameters, i.e. $R = R(p;x,t)$. For brevity, throughout this development, the dependence of R on x and t is not written out explicitly, although one must keep in mind that an output quantity is being calculated at each x and t of the model computation.

With given minimum p_{min}, most likely p_{likely}, and maximum p_{max}, values for p, set

$$\sin^2\theta_i = [p_{likely} - p_{min}]/(p_{max} - p_{mi})\qquad(A10)$$

for the ith component of p.

Suppose for the first approximation it is taken that $R(p_{max})$ and $R(p_{min})$ represent the 84% and 16% cumulative probability values, respectively. Then it is easy to show (Lerche et al., 1996) that

$$R(E_1(p)) \equiv E_1(R) = (1/2) \exp (\mu^2/2)[R(p_{max}) \exp (-\mu) + R(p_{min}) \exp(+\mu)]\qquad(A11)$$

and

$$\mu^2 = \ln \{1 + (1/2) \sum_{i=1}^{N} \alpha_i (R(p_{max}) - R(p_{min}))^2 \times$$

$$[R(p_{max}) (1 + 1/2 \exp (\mu^2/2-\mu)) + R(p_{min}) (1 + 1/2 \exp (\mu^2/2+\mu))]^{-2}\}\qquad(A12)$$

where $\alpha_i = \sin^4\theta_i + \cos^2\theta_i$.

In this way, it is possible to find estimates for the distributions of the set of assumption parameters at each grid-point and for every time-step using only two runs of the 2-D model; such a procedure saves considerable computing time.

Using these formulae one can find the mean and variance (μ^2) values at each grid-point (x) and at every time-step (t) for different outputs. Then at every time-step, t, for the output parameter j one can define the value of Global Variance

$$GVar_j(t) = \sum_{x} \mu^2_j(x,t)\qquad(A13)$$

where the sum is over all vector coordinates.

In the case of variation of only input parameters, the relative influence of this input on a given output can be defined as

$$RI_j(t) = GVar_j(t) / \sum_j GVar_j(t) \tag{A14}$$

⁸Breccia Hazards

A. General Observations

The mud volcanoes of the onshore and offshore regions of Azerbaijan are noted for their oil and gas accumulations on the flanks and crests of the volcanoes, making them commercially attractive as a major hydrocarbon province. Equally, however, the same mud volcanoes are noted for a variety of hazards which occur sporadically. These hazards range from sedimentary induced earthquakes, flame eruptions (to a height of up to 1.2 km), mud flows, explosive overpressure release, hydrate decomposition, and breccia ejection. Earlier in this volume estimates have been given for safety distances away from mud volcanoes for flame eruptions, for hydrate decomposition effects, for overpressure considerations, and for mud flow hazards in both aeolian and submarine conditions. In addition, earthquake hazard estimates based on the historical record have already been considered earlier.

The concern in this chapter is to estimate hazard distances for breccia ejecta from mud volcanoes in both aeolian and shallow submarine conditions. (Under the term "breccia" we include in this chapter pieces of rocks which are ejected with mud. In the literature one can be faced with the term "mud volcanic breccia", meaning often only the erupted mud). As a consequence of the historical observations, which indicate both direct breccia ejecta of rocks (up to a size of about $1m^3$) together with breccia ejected as clasts in mud flows, it would appear that a potential exists for aeolian rock ejection to distances of order a few kilometers for rocks up to 10 cm or so in linear size. For smaller pebbles the ejection distance can, it seems, reach further based on scattered observations.

Of interest here is precisely how rocks of a given mass distribution, emitted with velocity and angular distributions from a mud volcano, will produce a spatial distribution away from the ejection center. Also, their impact velocities are of interest from the hazard perception distances for pipelines, rigs and personnel. Clearly a volcano which is above the general level of the surrounding surface will have a greater hazard potential than a volcano whose crestal region is in a local depression relative to the surrounding surface (such as the Lokbatan mud volcano on the onshore Apsheron region of Azerbaijan which has been recorded as having exploded a minimum of 19 times over the last two centuries). And aeolian mud volcanoes can, presumably, eject breccia to further distances from their eruptive centers than similar mud volcanoes with submarine crestal regions although, depending on hydrate dissociation, such submarine mud volcanoes can present an equal or greater total hazard than aeolian mud volcanoes. For a submarine mud volcano, breccia constrained by the high viscous drag of the submarine environment then rapidly drop to the sea floor and can influence adversely submarine pipelines and subsea completion equipment. So the general problem of assessing safety distances for pipelines, equipment, rigs and personnel against rock ejecta is one of considerable practical importance.

Four factors provide dominant controls on the hazard problem. First is the height of the ejection region of a mud volcano above or below a surface of impact for the ejecta; second is the angle of the ejecta relative to the vertical - the direction of gravitation; third is the speed of ejection of breccia; and fourth is the viscous drag on a rock as a function of rock mass, area exposed to the viscous drag and the transport medium (water/air) viscosity.

Here we consider all four of these basic processes in relation to three different geological scenarios: aeolian transport from a mud volcano emitting breccia at a height above a surrounding plain; aeolian transport from a mud volcano sited in a depression and/or with a caldera lower than crater walls (such as the Lokbatan mud volcano of the onshore Azerbaijan region); submarine breccia transport in the absence of a flaming eruption.

Two major dynamical processes are included in the analysis; gravitation and viscous drag of the medium (air/water) on a rock fragment. The other two controlling influences on breccia ejecta are the initial release conditions of a rock mass and the geometric shape of the surrounding topography at the time of ejection relative to the release position of a rock mass.

The detailed mathematical analysis for each case is not difficult, but is tedious, and for that reason is recorded in the Appendices to this chapter.

B. Numerical Illustrations

In order to provide some typical order of magnitudes for dynamic range and impact speeds of ejected breccia, the case histories below are designed to provide estimates of the influence of various dynamical and geometrical factors in contributing to the hazard potential of breccia ejecta for rigs and platforms.

Six major groups of illustrations are given, split into two classes. The first class (of four groups) considers the ejection of breccia onshore with final impact onshore as well. The four groups considered in this class are: (i) planar emission and impact; (ii) emission from a high crater with impact on a lower plane; (iii) emission from a depressed crater with impact on a higher plane; (iv) emission from a walled crater with impact on a plane, and when the crater floor may be either depressed or elevated relative to the impact plane.

The second class (of two groupings) considers the ejection of breccia offshore. The two groupings considered here are: (i) emission from an island with impact on the ocean surface and ocean floor; (ii) emission from a submarine bank crater, which may also have an above-ocean crater wall.

As we shall see, the combined understanding of each of these separate case histories provides some rough and ready rules for estimating the potential hazards caused by breccia ejecta in both the onshore and offshore environments.

1. Class 1. Onshore Ejecta

a. Planar Emission and Impact

The simplest geometrical condition to consider is ejection of a rock mass in air from a point onto a horizontal plane passing through the same point. Two possible end-member viscosities are initially used to illustrate maximum ranges: molecular viscosity only, and a larger turbulent viscosity (see Appendices). Figures 8.1a,b show, respectively, the results of ejecting a 0.5 kg rock mass at an initial ejection speed of 50 m/s, and at 45° to the horizontal (the angle giving maximum range in the absence of viscosity) under molecular and turbulent viscosity, respectively. Note that the horizontal range is 254 m (molecular viscosity) and 170 m (turbulent viscosity) with corresponding impact speeds of 49 m/s and 44 m/s. The times to impact are 7.2s and 6.4s, respectively, after ejection. Thus, one has a very short time after ejection before being impacted at a distance of between 170-250m away from the ejection point.

For completeness it is relevant to note that in the presence of turbulent viscosity the maximum horizontal range is <u>not</u> achieved at the emission angle of 45° (the vacuum condition) but rather at about 42° as shown in figure 8.1c, when the maximum range is 170.9m, and a flight time of 6.08s.

Because the turbulent viscosity depends on rock size, different mass rocks (even when arranged to be thrown at angles giving maximum range) will have different maximum ranges. Thus figure 8.2a shows the result of ejecting a rock mass of 5×10^{-3} kg at its maximum range angle of 36° and at 50 m/s, yielding a range of 85.9 m; while a similar rock of 5×10^{-2} kg has an angle of 40° and a maximum range of 127.7 m. Correspondingly, as shown in figures 8.2c and 8.2d, rock masses of 5 kg and 50 kg have angles of 44° and ranges of 206.4 m and 230 m, respectively. The corresponding impact speed (starting each rock mass at 50 m/s) is low (29 m/s) for low mass (0.005 kg) rocks rising to very close to the initial ejection speed of 50 m/s for high mass (≥ 5 kg) rocks, reflecting the larger role of turbulent viscosity in reducing the smaller mass particles to the limiting terminal speed more quickly compared to the larger mass ejecta.

The maximum range of an ejected particle also depends sensitively on its initial emission angle. Shown in figures 8.3a-d are the results for emission of a 0.5 kg rock at initial speed of 50 m/s, for emission angles of 85°, 65°, 25° and 15° to the horizontal, respectively, yielding the corresponding ranges of 30m, 80m, 145m, and 104m, respectively, with impact speeds of between 50-44 m/s, and the lower speeds referring to lower angles of emission.

Thus: typical rock breccia will have ranges of 30-250 m, impact speeds of around 30-50 m/s (when started at 50 m/s) and take about 5-8s to impact. The range of angles yielding maximum horizontal ranges of greater than 100 m is between about 65°-15° to the horizontal.

As the initial ejection speed is varied so, too, the maximum horizontal range and flight time vary. Thus figures 8.4a and 8.4b show the results for a

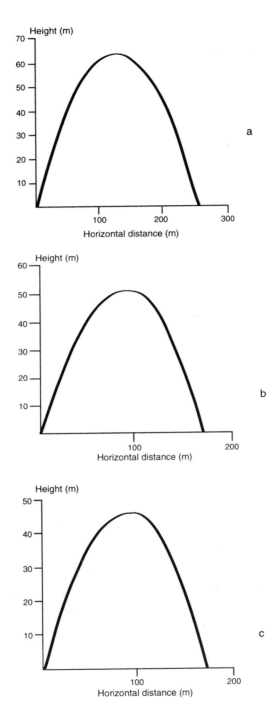

Figure 8.1. Results of ejecting a 0.5 kg rock mass in air at 50 m/s: (a) at 45° with molecular viscosity; (b) at 45° with turbulent viscosity; (c) at 42°, the largest range angle under turbulent viscosity.

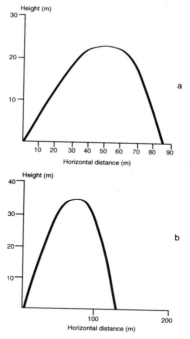

Figure 8.2. Different mass rocks ejected at 50 m/s under turbulent viscosity: (a) 5 x 10⁻³ kg at 36°; (b) 5 x 10⁻² kg at 40°. The different angles are maximum range angles.

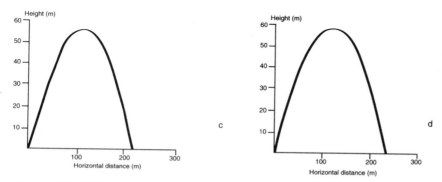

Figure 8.2. Different mass rocks ejected at 50 m/s under turbulent viscosity: (c) 5 kg at 44°; (d) 50 kg at 44°. The different angles are maximum range angles.

222

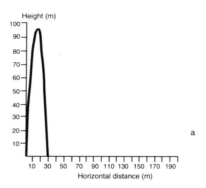

Height (m)

Horizontal distance (m)

a

Figure 8.3. Ejection of a 0.5 kg rock mass in air at 50 m/s under turbulent viscosity for emission angles of: (a) 85°.

0.5 kg rock thrown at 45° to the horizontal at speeds of 5 m/s (figure 8.4a) and 500 m/s (figure 8.4b), which values bracket expected extremes of ejection. The results are ranges of between 2.5m and nearly 900m, and corresponding flight times of 0.8s to 19s.

Thus the hazard occurs quickly (timescale less than around 20s) and with a maximum impact speed less than about 65 m/s (in the case of ejection at 500 m/s). One might then estimate a worst case hazard distance of 1 km, a worst case impact speed of 100 m/s, a worst case impact mass of 5-50 kg and a time of less than 20s.

b. Raised Crater and Planar Impact

Breccia ejected from a crater raised higher than the impact plane can not only travel further than similar ejecta from a crater on the impact plane, but can also impact with higher velocities than their initial ejection speeds - something which is not possible for breccia ejected on the impact plane. For instance, shown on figures 8.5a-c are the results for breccia ejection from a 100 m high volcano for rock masses of 0.05kg, 0.5kg and 5kg, respectively, emitted at 50m/s. The corresponding lateral impact distances are 146m, 231m, and 278m, and corresponding impact speeds are 39 m/s, 52m/s, and 68m/s.

The same range of rock masses, when ejecta from a 400m high crater, impact at 217m (for 0.05kg), 317m (for 0.5kg) and 402m (for 5kg), with corresponding impact speeds of 39m/s, 55m/s and 73 m/s as shown in figures 8.6a,b, and c, respectively.

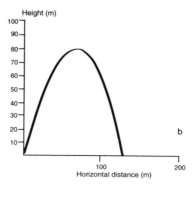

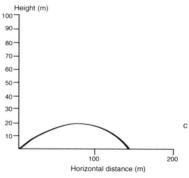

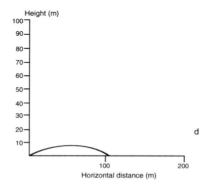

Figure 8.3. Ejection of a 0.5 kg rock mass in air at 50 m/s under turbulent viscosity for emission angles of: (b) 65°; (c) 25°; (d) 15°.

224

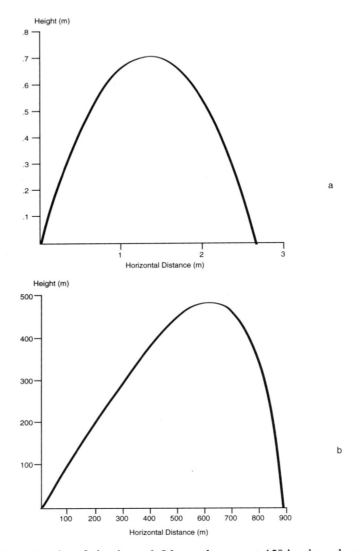

Figure 8.4. Results of ejecting a 0.5 kg rock mass at 45° in air under turbulent viscosity at different speeds: (a) 5 m/s; (b) 500 m/s.

While the lateral range is up to 35% higher for ejection from the 400m high crater compared to the 100m high crater, there is little change in the impact speed because the turbulent viscosity is causing all rock masses to be close to their terminal velocities. Typical transit times from ejection to impact are between 10-15 seconds.

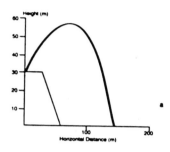

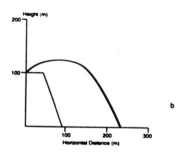

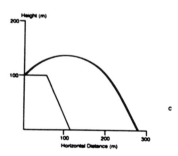

Figure 8.5. Breccia ejection from a 100 m high crater at 50 m/s for rock masses of: (a) 0.05 kg; (b) 0.5 kg; (c) 5 kg.

Thus, a worst case hazard would suggest a range of around 300-400m for ejection from a mud volcano crater of order 100-400m high, and an impact speed of between about 50-100m/s.

226

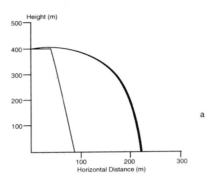

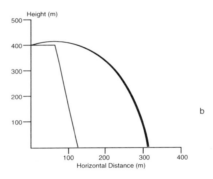

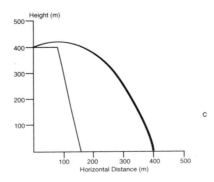

Figure 8.6. As for figure 8.5 but with the crater height at 400 m above the impact plane.

c. Walled Crater Ejection and Planar Impact

There are two obvious sorts of end-member walled craters that can be considered: a crater whose floor is lower than the impact plane; and a crater with a higher floor than the impact plane. In either case, however, the ejected breccia must have trajectories that are higher than the wall in order to end up on the impact plane. This limitation provides a minimum angle for ejection in order to pass over the wall on the rise phase of the trajectory, and also a maximum ejection angle such that on the descent phase of the trajectory the breccia will end up outside the walled crater.

Figures 8.7a,b,c and d show the lowest angle trajectories for breccia ejected at 50 m/s and with masses of 0.05kg, 0.5kg, 5kg and 50kg, respectively, for a crater floor 50m lower than the wall height and also 20m below the impact plane. The corresponding impact ranges are from 104m (0.05kg) through to 207m (50kg).

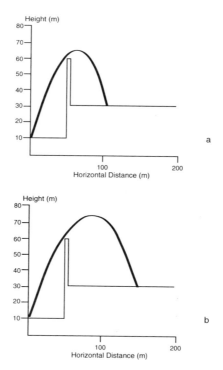

Figure 8.7. Lowest angle trajectories for breccia ejected at 50 m/s and with masses of (a) 0.05 kg; (b) 0.5 kg.

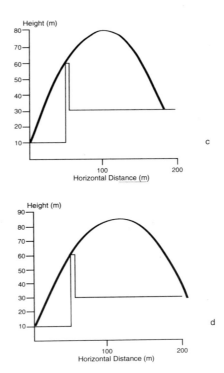

Figure 8.7. Lowest angle trajectories for breccia ejected at 50 m/s and with masses of (c) 5 kg, for a crater floor 50 m lower than the wall height and also 20 m below the impact plane; (d) 50 kg, for a crater floor 50 m lower than the wall height and also 20 m below the impact plane.

The upper limiting angle is illustrated in figures 8.8a and 8.8b for a 0.5kg rock ejected at 50 m/s from the same depressed crater floor. An initial angle of 81° to the horizontal (figure 8.8a) results in the rock impacting the inner part of the crater wall on its descent, while a 70° angle of emission enables the rock to just miss the crater wall and end up on the impact plane at 59 m from the ejection point. The lower critical angle for emission in this case is 53° as shown in figure 8.8c, when the 0.5kg rock just grazes the crater wall and ends up on the impact plane at 149m from the ejection point.

For a raised crater floor, 20m higher than the impact plane, and a less pronounced crater rim height (20m from the crater floor), a wider set of emission angles will allow the breccia to end up on the impact plane, as shown in figures 8.9a,b, and c, respectively.

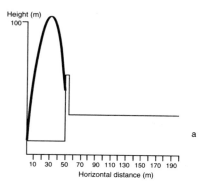

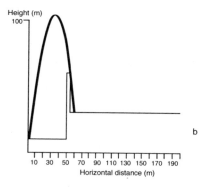

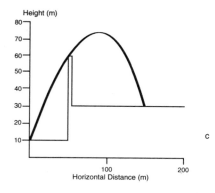

Figure 8.8. Upper limiting angle effect for a 0.5 kg rock mass ejected at 50 m/s from the same depressed crater of figure 8.7: (a) 81°; (b) 70°; (c) 53°.

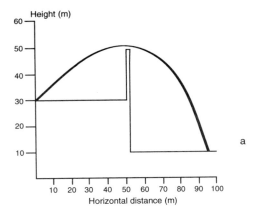

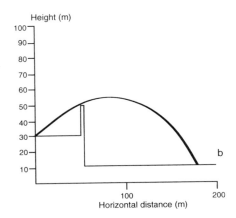

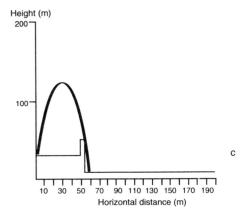

Figure 8.9. A raised crater floor, 20 m higher than the impact plane, and a crater rim of only 20 m from the crater floor give a wider set of initial angles allowing impact: (a) 34°; (b) 29°; (c) 80°.

d. Depressed Crater Emission onto a Higher Impact Plane

This situation is most easily modeled by supposing the crater wall of the previous sub-section is of infinite width. Again there are two critical angles of emission such that breccia outside the upper and lower critical angles remain within the crater. This case is exhibited in figures 8.10a,b and c for a rock mass of 0.5kg ejected at 50m/s. Figure 8.10a shows the upper critical angle of 72°; figure 8.10b the lower critical angle of 32°; while figure 8.10c shows the angle of 48° providing maximum range; and the range is then 144m. Figures 8.11a and b show how the maximum range increases as rock mass increases, from 178m for a mass of 5kg to 200m for a 50kg mass, all other parameters being held fixed.

Likewise, if the ejection speed of the mass is increased then so, too, is the range increased. For a mass of 0.5kg, figures 8.12a and 8.12b show a maximum range of 360m for ejection at 100m/s and 980m for ejection at 500m/s. The corresponding impact velocities in these two cases are 65 m/s and 89 m/s, reflecting the terminal drag velocities.

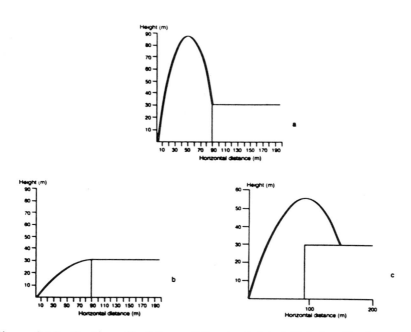

Figure 8.10. Crater wall of figure 8.9 extended indefinitely. Rock mass of 0.5 kg is ejected at 50 m/s. Upper and lower angles of emission are: (a) 72°; (b) 32°; while angle giving maximum range is 48° (c).

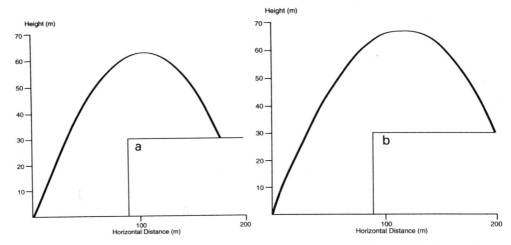

Figure 8.11. Maximum range for emission from crater of figure 8.9 when rock mass increases: (a) 5 kg; (b) 50 kg; ejection at 50 m/s.

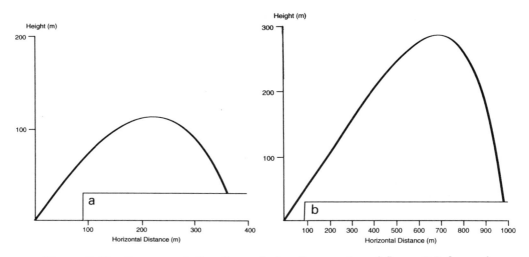

Figure 8.12. Range variation for emission from crater of figure 8.9 for rock mass of 0.5 kg when ejection speed increases: (a) 100 m/s; (b) 500 m/s.

Thus, again, range scales are of order several hundred meters, with impact speeds of around 30-100m/s, and flight times of a few seconds to about 20 seconds.

Summarizing the general rough values for Case 1 type emissions: under just about all conditions a range of around 500±500m is appropriate, with impact speeds of 10-100m/s for rock masses between 0.05kg-50kg, with higher speeds of impact for the higher mass rocks.

The effects of a raised crater tends to increase range and impact speed while a depressed crater does precisely the opposite. Timescales after emission tend to be in the few second to around 20 second range, leaving little time for hazard preparation.

The danger zone is clearly within about 1km of breccia eruption, and a worst hazard is a 50kg rock arriving at of order 100m/s; such a rock impact could cause serious rig damage.

2. Class 2. Ejecta under Offshore Conditions

a. Island Emission into the Ocean

The mud islands that appear and disappear in the South Caspian Sea do so irregularly, and are known to provide explosive eruptions of breccia, mud, gas, and flames at random. Most often the islands range from being at or just above sea level to being several meters to tens of meters above sea-level (see later in this volume).

But, in either event, the hazard prognosis is clearly two-fold: first is the maximum range for an ejected rock to reach the sea-surface, and second is the impact of such a rock not only at the ocean surface, where rigs can be adversely affected, but also on the ocean floor where sub-sea completion equipment and submarine pipelines can be influenced. The complexity here is caused by the two media in which a rock is transported: first in air (with low viscosity) after ejection from the island eruption and then in the ocean with much higher viscosity.

The air transport fraction of the flight path is identical to the paths just described for aeolian transport. The exception here is that once the rock has "impacted" on the sea-surface plane it now sinks to cause an extra hazard.

Thus, at the sea-surface the worst case hazard values for surficial equipment are again as given previously. The sub-sea hazards are now determined by the impact of the rocks on the ocean floor. For instance, as shown in figure 8.13a, a 0.5kg rock ejected at 50 m/s has an air range of around 170m when it impacts the ocean surface. Thereafter its ocean bottom terminal velocity is 2m/s in a 50m deep ocean with a total flight time of around 32 secs. The same rock in a 5m deep ocean also reaches a terminal speed of 2m/s with a total flight time of just over 9 sec. A lower mass (0.05kg) rock has terminal speed of 1m/s and a total range of 135m as shown in figure 8.13c, while a 5 kg (50kg) rock has a range of 218m (245m) and a terminal water bottom speed of 3m/s (4m/s), as shown in figure 8.13d (8.13e),

234

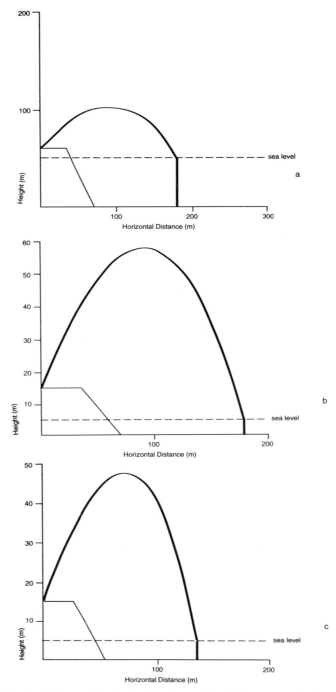

Figure 8.13. A 0.5 kg rock ejected in air at 50 m/s has a terminal speed of: (a) 2 m/s in a 50 m deep ocean and takes 32 s transit time; (b) 2 m/s in a 5 m deep ocean but with a transit time of 9 s. A lower mass (0.05 kg) rock has a terminal speed of 1 m/s (Figure 8.13c).

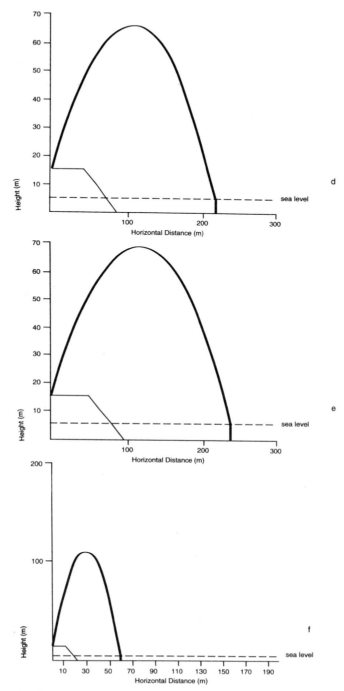

Figure 8.13. A 5 kg rock has a range of 218 m and a terminal speed of 3 m/s (figure 8.13d); while a 50 kg has a range of 245 m and a terminal speed of 4 m/s (figure 8.13e).

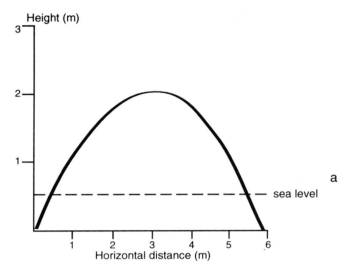

Figure 8.14. Submarine ejection of a rock mass of: (a) 0.5 kg at 50 m/s in 0.5 m water at maximum range emission angle of 50°.

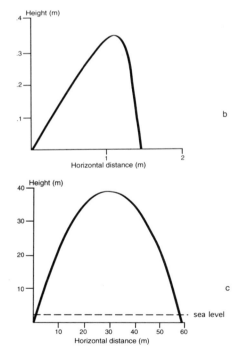

Figure 8.14. Submarine ejection of a rock mass of: (b) 0.5 kg rock emitted at 20°; (c) a 5 kg rock at 500 m/s.

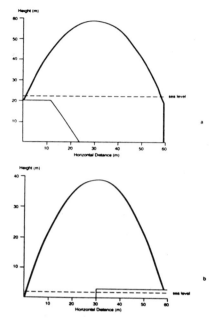

Figure 8.15. Ejection from a shallow submarine bank: (a) deep water (22 m) impact of a 5 kg rock emitted at 500 m/s; (b) same rock conditions but now to a shallow island impact.

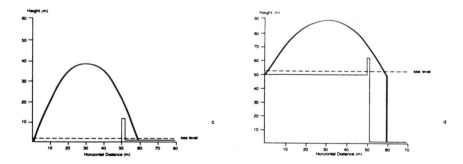

Figure 8.15. Ejection from a shallow submarine bank: (c) submarine emission (with an aeolian crater wall) to impact on another shallow submarine bank; (d) emission past an exposed crater wall to deep (52 m) water impact.

Indeed, other similar calculations show that once the water depth is greater than about 2m, almost all case histories reach a terminal water speed of 1-4m/s. Shallower water depths can be well approximated by impact at the ocean surface for hazard conditions.

b. *Submarine Emission*

In this case, as for breccia emission from a submarine bank eruption, the initial flight path after emission is through the overlying water. If the water depth is greater than about a meter or two, then it is extremely difficult for ejected breccia to punch through the water into the overlying air. For instance, shown on figure 8.14a is the case of a 0.5kg rock emitted into 0.5m of water at 50m/s. The emission angle giving maximum height (and so penetrating the overlying water column) is 50°. If that angle is reduced to 20°, the same rock never gets out of the overlying water column and rises only 0.35m above emission depth before falling back, as shown in figure 8.14b. Of course, by increasing the rock mass to 5kg and its ejection speed to 500 m/s it is possible to drive the rock through the overlying water column into air, as shown in figure 8.14c, with a height of only 39m, because considerable energy is expended in overcoming the turbulent water friction.

Similar patterns of motion are available for shallow submarine bank ejecta into deep water or with island craters surrounding the submarine bank emission point. Figures 8.15a-d show four such cases, with figure 8.15a showing deep water (22m) impact at 3 m/s for a 5kg rock ejected at 500 m/s from a shallow submarine bank. Figure 8.15b shows the same ejection but now from the submarine bank to a shallow island. Figure 8.15c shows the complications caused by submarine emission being required to pass above an aeolian crater wall and land on another shallow, submerged bank; while figure 8.15d shows emission from a shallow submarine bank past an aeolian exposed crater wall to deep (52 m) water.

In all these cases, and others that have been investigated, simple breccia emission in submarine conditions require shallow water ($\lesssim$2m) to have a significant lateral range. Of course, accompanying breccia emission with mud flows or island raising, so that the overlying water is not a problem, helps with dynamic lateral range.

Otherwise submarine emission, even for high velocity (~500 m/s) rocks, does not have a dynamic range of more than about 100 m and an ocean bottom terminal speed of 1-4m/s. This result is to be contrasted with emission from islands (similar to onshore emission) where lateral ranges of order several hundred meters are possible with ocean surface impact speeds of up to around 100 m/s and ocean bottom impact speeds of 1-4 m/s.

C. Conclusions

The lessons to be learnt from this series of investigations, in concert with the observed distributions of breccia onshore, is that aeolian emission of rocks is the most hazardous condition, irrespective of whether the impacting domain is terrigenous or marine.

As a general rule, seconds to tens of seconds are enough to create a hazard timescale after breccia ejecta, with hazard distances typically to order a few hundred meters, but occasionally reaching kilometer for largish (5kg) rocks emitted at high (~500 m/s) speeds.

In the impact zone, velocities of impact are in the general range of 10-100 m/s in air or at an air-water interface, and 1-4 m/s in water at the water base for water depths greater than about 1-2 m. For shallower water depths, the air-water impact speed is a good approximation.

This hazard is, perhaps, one that should be of some considerable concern to onshore and offshore equipment within about 0.5-1 km from breccia ejection points, and also one of concern to any subsea completion equipment (e.g. pipelines) over the same lateral hazard distance.

Appendix A

A. Single Rock Fragment Motion in Air

Consider a rock of mass m, ejected from a mud volcano at speed V_0, and with initial angle θ_0 to the horizontal as sketched in figure A1. Let the ejection point be at a height h above a reference plane. The viscous drag on the mass due to air friction depends on the shape of the mass. For instance, for a spherical mass of radius r, the viscous retarding force is $-6\pi\eta r v$, where η is the coefficient of viscosity which is taken to be constant for the purposes of illustration (A velocity-dependent viscosity is considered later in this Appendix). For an irregularly shaped mass one customarily then writes a viscous force as $-\lambda\eta L v$ where λ is a coefficient dependent on the shape of the mass, and L is an effective length scale of the mass, and v is the particle velocity.

Under a vertical gravitational retardation of acceleration g, the equations of motion of the mass are

$$m\ddot{z} = -mg - \lambda\eta L|\dot{z}| \qquad (A1)$$

and

$$m\ddot{x} = -\lambda\eta L|\dot{x}| \qquad (A2)$$

for vertical(z) and horizontal(x) components of motion, respectively.

For a mass starting at time t=0 at position z=h, $x=x_0$, and with initial velocity components $\dot{z} = V_0\sin\theta_0$, $\dot{x} = V_0\cos\theta_0$, the general solution to equations (A1) and (A2) is

240

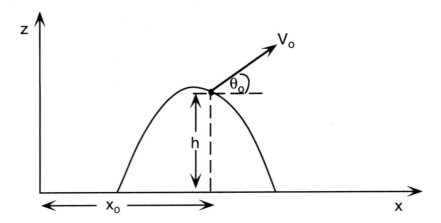

Figure A1. Sketch of the parameters defining particle motion.

$$x(t) = x_0 + \tau V_0 \cos\theta_o (1 - \exp(-t/\tau)) \qquad (A3a)$$

and

$$z(t) = h - g\tau t + \tau(g\tau + V_0 \sin\theta_0)(1 - \exp(-t/\tau)) \qquad (A3b)$$

where $\tau = m/(\lambda \eta L)$.

Note that the corresponding velocity components are

$$\dot{x}(t) = V_0 \cos\theta_0 \exp(-t/\tau) \qquad (A4a)$$

$$\dot{z}(t) = -g\tau + (g\tau + V_0 \sin\theta_0) \exp(-t/\tau) \qquad (A4b)$$

so that the total speed is

$$\begin{aligned} v &= (\dot{x}^2 + \dot{z}^2)^{1/2} \\ &= [V_0^2 \exp(-2t/\tau) - 2g\tau V_0 \sin\theta_0 \exp(-t/\tau)(1 - \exp(-t/\tau)) \\ &\quad + g^2\tau^2 (1 - \exp(-t/\tau))^2]^{1/2} \qquad (A5) \end{aligned}$$

which, as $t \to \infty$, reduces to the terminal speed $v = g\tau$.
The maximum height reached occurs when $\dot{z} = 0$, i.e. at a time t_0 given by

$$t_0 = \tau \ln\{1 + V_0 \sin\theta_0/(g\tau)\}$$

and at time t_0, one has

$$z(t_0) = h + V_0 \sin\theta_0 \tau - g\tau^2 \ln\{1 + V_0 \sin\theta_0/(g\tau)\} \quad (A6a)$$

and

$$x(t_0) = x_0 + \tau V_0^2 \sin\theta_0 \cos\theta_0/(g\tau + V_0 \sin\theta_0) \qquad (A6b)$$

A given plane $z = z_1$ is crossed by a particle at a time $t = t_1$, where t_1 is implicitly given by

$$t_1 = (g\tau)^{-1} \{h - z_1 + \tau(g\tau + V_0\sin\theta_0)(1 - \exp(-t_1/\tau))\}$$
(A7a)

and the lateral distance covered at $t = t_1$ is then

$$x(t_1) = x_0 + \tau V_0\cos\theta_0(1 - \exp(-t_1/\tau))$$
(A7b)

Because one requires $t_1 \geq 0$, it follows from equation (A7a) that a plane $z = z_1$ can only be crossed when

$$z_1 \leq h + \tau(g\tau + V_0\sin\theta_0)(1 - \exp(-t_1/\tau))$$
(A8)

The right side of inequality (A8) is certainly less than the maximum value of $h + \tau(g\tau + V_0\sin\theta_0)$, which sets an upper limit. This factor will become of significance in the cases of breccia ejecta from a mud volcano in a depression and also in a submarine setting.

Note also that the angle, θ_{max}, which gives the largest horizontal distance traveled, x_{max}, to reach a plane $z = z_1$ is given through

$$\ln(1 + \varepsilon\csc\theta_{max}) = (h - z_1)/(g\tau^2)$$
$$+ \varepsilon\csc\theta_{max}(1 + \varepsilon\sin\theta_{max})/(1 + \varepsilon\csc\theta_{max})$$
(A9)

where $\varepsilon = V_0/g\tau$, which (when computed for θ_{max}), can then be used in equations (A7a) and (A7b) to obtain x_{max} and the time, $t_{1,max}$, after ejection to reach both $z = z_1$, and

$x = x_{max}$.

Note, for later convenience, that if the viscosity is very small, so that the decay time-scale τ is large compared to any other time-scale of interest, then equation (A9) yields the approximate solution

$$\sin\theta_{max} = 2^{-1/2}(1 + g(h - z_1)/V_0^2)^{-1/2}$$
(A10)

To this point the analysis of the basic dynamical equations is completely general. However, the intervention of topographic distortions away from a plane and/or of a change in the viscosity of the transport medium (water to air in the case of submarine ejection for those rocks that have enough speed to transit the water-air boundary) makes the final spatial distribution of ejected rock fragments and their impact speeds more complex than indicated by the general free-space solutions given above.

It is for these reasons that we now first spell out the restricted motions of ejected breccia for the three geometric/transport medium cases.

B. Breccia Ejection in Air from a Walled Crater

Consider a walled mud volcano as sketched in figure A2 with a wall of height Δ above the breccia ejection height, h, as sketched. Ejection is taken to occur from within the crater as shown on figure A2. Then an ejected rock fragment

242

(ejected at position $x=x_0$, $z=h$, at time $t=0$, with initial angle θ_0 and speed V_0) can only escape from the crater if, when it reaches $x=L+D$ ($> x_0$), it has attained a height value of $z \geq \Delta+h$. All other rock fragments are contained within the crater. (It is sufficient to consider only rocks ejected with positive angles, $\pi/2 \geq \theta_0 \geq 0$, because, as we will see in Appendix B, one can rotate the coordinate system to cover three-dimensions).

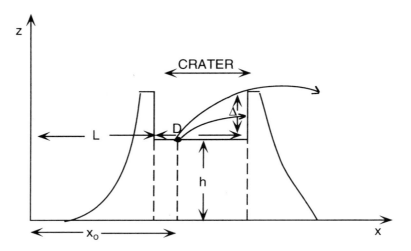

Figure A2. Sketch of parameters defining emission from a raised crater past a crater wall to impact on a lower plane.

Then, from equation (A3a) one has $x=L+D$ at a time t_2 given by

$$t_2 = \tau \ln \left\{ \frac{\tau V_0 \cos\theta_0}{[\tau V_0 \cos\theta_0 + x_0 - (L+D)]} \right\} \qquad (A11)$$

and t_2 is positive only when

$$L + D \geq x_0 \qquad (A12a)$$

and when

$$\tau V_0 \cos\theta_0 > L+D - x_0 \geq 0. \qquad (A12b)$$

All other cases have the ejected rock hitting the crater floor before it reaches the crater wall. For those rocks for which (A11) is satisfied, subject to (A12a) and (A12b) being in force, the vertical location of the rock at time t_2 is

$$z(t_2) = h - g\tau t_2 + (\tau g + V_0 \sin\theta) \frac{(L+D-x_0)}{V_0 \cos\theta_0} \qquad (A13)$$

But we require $z(t_2) \geq h+\Delta$ if a rock fragment is to escape the crater. Then write this escape condition as

$$(\tau g + V_0\sin\theta_0)(L+D-x_0) \geq V_0\cos\theta_0\left[\Delta + g\tau^2\ln\left\{\frac{\tau V_0\cos\theta_0}{(\tau V_0\cos\theta_0 + x_0 - (L+D))}\right\}\right]$$

(A14)

and subject also to inequalities (A12a) and (A12b).

Inequality (A14) provides the angular and initial velocity controls for the distribution of rock fragments escaping from the crater once ejected from a lateral position x_0 inside the crater (i.e. $L \leq x_0 \leq L+D$).

C. Breccia Ejecta from a Depressed Crater

Many of the Azerbaijani onshore mud volcanoes sit in relatively shallow depressions relative to the surrounding terrain. Ejections of breccia also take place from such craters. To provide a model behavior take the depression (without a wall around the crater) to be of the symmetric shape

$$z_{surface}(x) = z_* - \Delta\operatorname{sech}^2(x/L)$$

(A15)

where the depression center is at a depth Δ relative to the terrain and extends laterally over a scale L across the center. As we shall show in the next Appendix, the values of Δ and L can be generalized to allow for the three dimensional nature of the depression. In this Appendix they are taken to be constant in the line of ejection of the breccia in the x-direction.

The distance traveled laterally by an ejected rock fragment then depends on its ejection point relative to the central depression minimum and upon its initial angle of ejection as sketched in Figure A3. The slope of the surface $z_{surface}(x)$ can be written in terms of an equivalent, spatially variable, angle $\theta_{sur}(x)$ given through

$$\tan\theta_{sur} = -2(\Delta/L)\tanh(x/L)\operatorname{sech}^2(x/L)$$

(A16)

and only rock fragments can be ejected with initial angles θ_0 lying between $\pi+\theta_{sur} \geq \theta_0 \geq \theta_{sur}$ for positive values of θ_{sur}, and between $\pi-|\theta_{sur}| \geq \theta_0 \geq -|\theta_{sur}|$ for negative values of θ_{sur}.

With that restriction understood, the final position reached by a rock fragment is obtained by substituting equation (A3a) for x(t) and equation (A3b) for z(t) into equation (A15) (i.e. replace $z_{surface}$ by z(t), and x by x(t)). Then one solves the resulting equation for the time t_* (≥ 0) as a function of initial location of the rock

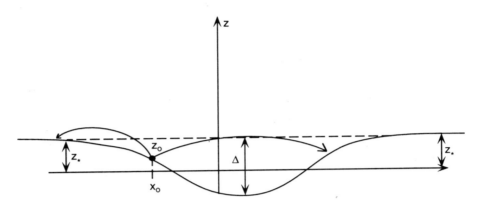

Figure A3. Sketch of parameters for emission from a depressed crater.

sample on the surface (i.e. at time t=0 with $x = x_0$ the initial vertical location is $z_0 = z_* - \Delta sech^2(x_0/L)$). In this way, once the time t_* is determined, both the final location and final impact speed of the rock fragment on the surface are determined.

D. Submarine Breccia Ejection

In the case of submarine breccia eruptions from mud volcanoes there are both dynamical complications in the motion of breccia and also double hazards due to drilling rigs and production platforms being sited on the ocean surface, and due to submarine pipelines and subsea completion equipment being located on the ocean floor. For rigs and platforms the hazard arises from ejected breccia which have sufficient energy to transit to the ocean surface from the submarine ejection point and, thereafter, are moving in air to re-enter the ocean. Rigs or platforms in the paths of such rock fragments are then subject to hazardous bombardment. For rock fragments which either remain completely in the ocean after ejection or which miss the rig or platform and so re-enter the ocean, they eventually settle to the ocean bottom and can crush or damage pipelines and subsea completion equipment. Hence the double hazard statement above.

To provide a model of this process consider emission from a mud volcano whose crest is at a depth D below the ocean surface, and with emission from a location z=h, x=x₀ above the ocean floor (taken as planar and in the origin plane of the coordinate system as sketched in figure A4). The dashed lines in figure A4 represent schematically the two sorts of particle paths available. The major difference in this situation compared to the previous cases is in the massive contrast between water viscosity, η_w, and air viscosity, η_a, which is reflected in two different viscous time-scales τ_w and τ_a.

For particles which stay wholey submerged, precisely the same analysis can be repeated as for the previous aeolian situations described (with the replacement of τ by τ_w) because such particles are always in a single medium.

But for particles which can reach the plane $z = h+D$ at any x-coordinate, they then find themselves in air and move according to different trajectories due to the much higher value of τ_a compared to τ_w. To determine which particles can reach $z=h+D$

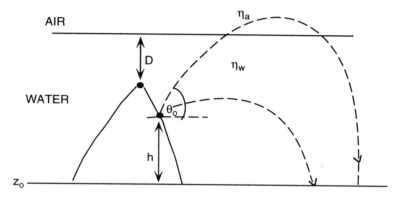

Figure A4. Sketch of trajectories for submarine emission of breccia.

after submarine ejection at angle θ_0 with initial speed V_0 from the submarine location $z=h$, $x=x_0$, is relatively simple.

Use equation (A3a) (with τ replaced by τ_w) and insert $z=h+D$. Then the time $t_3(\geq 0)$ at which a particle reaches the ocean surface after emission is given by the solution (for t_3) to

$$D = -g\tau_w t_3 + \tau_w(g\tau_w + V_0\sin\theta_0)(1-\exp(-t_3/\tau_w)) \qquad (A17)$$

provided a positive solution exists. Clearly if $D=0$, inspection of equation (A17) shows that a positive t_3 exists. But as D increases eventually there fails to be a solution to equation (A17) for specified V_0, θ_0 and τ_w. Indeed, there fails to be a solution when $\dot{z} \leq 0$ on $z = D+h$, which occurs when

$$t_3 \geq \tau_w \ln\{1 + V_0\sin\theta_0/g\tau_w\} = t_* \qquad (A18a)$$

i.e. when

$$D \geq g\tau_w V_0\sin\theta_0 - g\tau_w^2 \ln\left(1 + \frac{V_0\sin\theta_0}{g\tau_w}\right) = D_{crit} \qquad (A18b)$$

there is no particle emitted with V_0 at θ_0 from depth D which will reach the ocean surface.

For particles <u>not</u> obeying inequality (A18a) or inequality (A18b) they <u>do</u> reach the ocean surface at time t_3, with z-coordinate D+h; and with x-coordinate given by

$$x_w(t_3) = x_0 + \tau_w V_0 \cos\theta_0 \ (1-\exp(-t_3/\tau_w)) \quad\quad (A19a)$$

and velocity components

$$\dot{x}(t_3) = V_0\cos\theta_0 \ \exp(-t_3/\tau_w) \quad\quad (A19b)$$

$$\dot{z}(t_3) = -g\tau_w + (g\tau_w + V_0\sin\theta_0) \ \exp(-t_3/\tau_w) \quad\quad (A19c)$$

For times, t, in excess of t_3 such particles are then in air, with viscous decay constant τ_a. Thus their trajectories in air (for $t{\geq}t_3$) are given by

$$x(t) = x_w(t_3) + \tau_a \ \dot{x}(t_3) \ (1-\exp(-(t-t_3)/\tau_a)) \quad\quad (A20a)$$

$$z(t) = z_w(t_3) - g\tau_a(t-t_3) + \tau_a \ (g\tau_a + \dot{z}(t_3))(1- \exp(-(t-t_3)/\tau_a)) \quad\quad (A20b)$$

and these trajectories are appropriate until, once again, the rock fragment reaches the ocean surface, z=D+h, at time $t{=}t_4{>}t_3$, given implicitly by

$$g\tau_a(t_4{-}t_3) = \tau_a(g\tau_a + \dot{z}(t_3))(1- \exp(-(t_4{-}t_3)/\tau_a)) \quad\quad (A21)$$

Such particles re-enter the water at coordinates $x(t_4)$ and $z(t_4) = D{+}h$, with corresponding velocity components

$$\dot{x}(t_4) = \dot{x}(t_3) \ \exp(-(t_4{-}t_3)/\tau_a) \quad\quad (A22a)$$

$$\dot{z}(t_4) = -g\tau + (g\tau + \dot{z}(t_3)) \ \exp(-(t_4{-}t_3)/\tau_a) \quad\quad (A22b)$$

Thereafter, the particles sink in the ocean, finally reaching the ocean floor (z=0) at time t_5 given through

$$t_5 = t_4 + (g\tau_w)^{-1}\left\{h + \tau_w(g\tau_w + \dot{z}(t_4)) \left(1{-}\exp\left(-(t_5{-}t_4)/\tau_w\right) \right) \right\} \quad (A23a)$$

and at x-coordinate

$$x(t_5) = x(t_4) + \tau_w \dot{x}(t_4)(1- \exp(-(t_5{-}t_4)/\tau_w)) \quad\quad (A23b)$$

and with velocity components

$$\dot{x}(t_5) = \dot{x}(t_4) \ \exp(-(t_5{-}t_4)/\tau_w) \quad\quad (A23c)$$

$$\dot{z}(t_5) = -g\tau_w + (g\tau_w + \dot{z}(t_4)) \ \exp(-(t_5{-}t_4)/\tau_w) \quad\quad (A23d)$$

One has to be careful in deciding on the viscosity to use. The classical (Lamb, 1879) viscosity, η_c, is only appropriate when the speed, v, is so small that $\Re \equiv |v|a/(v_c \, 6\pi) << 1$, where v_c is the kinematic viscosity ($v_c = \eta_c/\rho$), ρ is the air or water medium density, and a is the linear dimension scale of the thrown object. When $\Re \gtrsim 1$, a more relevant model of viscosity is (Lamb, 1879)

$$\eta = \eta_c \, (\Re + 1).$$

Because the particle speed, $|v(t)|$, depends on time after ejection, so too does η. However, this is not a significant problem numerically where time steps Δt are used, because one uses the viscosity η at time-step $(n\Delta t)$ to update the velocity at time $(n+1)\Delta t$. Accordingly, in each time-step interval one can use a constant viscosity based on the previous time-step of calculation. Note that once the particle speed drops to the point where $\Re << 1$, then one recovers classical motion under molecular viscosity behavior of the transporting medium.

Thus one has a complete traceable history of each particle as it moves from its original coordinates to its final resting place.

In this case, write $u=\dot{x}$ and $v=\dot{z}$ when equations (A1) and (A2) can be written

$$\dot{v} = -g - \frac{|v|}{\tau(u,v)}$$

$$\dot{u} = -\frac{|u|}{\tau(u,v)}$$

where $\tau = m/(\lambda L \, \eta_c(\Re(u,v) + 1))$ so that a numerical scheme for evaluating motion of a particle is given through

$$v_{n+1} = \{-g - \frac{1}{\tau(u_n,v_n)}|v_n|\} \, \Delta t + v_n$$

$$u_{n+1} = -[|u_n|/\tau(u_n,v_n)] \, \Delta t + u_n$$

$$z_{n+1} = z_n + v_n \, \Delta t$$

$$x_{n+1} = x_n + u_n \, \Delta t$$

where Δt is the time-step chosen and the initial conditions for velocity and position of the particle are as described earlier.

The time-step, Δt, is free to be chosen. However, the spatial step, Δl, can hardly be any smaller than the size of the mass being ejected and, when turbulent viscosity is allowed for, must be several times the object size, L.

Then set $\Delta l = mL$, where m is a number larger than unity. Then at step n one has

$$\Delta t = \Delta l/(u_n{}^2 + v_n{}^2)^{1/2}.$$

This numerical procedure is extremely stable, accurate, and fast numerically. It also honors all of the essence of the prescriptions and discussions given previously.

E. Initial Distribution Considerations

1. Initial Mass Distributions

Examinations of breccia in Azerbaijan show large rock fragments (upwards of a meter or so in linear dimensions) often at several kilometers from an eruption site. However, care has to be exercised when one attempts to determine mass distributions including such large but distant samples. The reason, of course, is that breccia are also carried by mud flows and, after rain washing away the mud, all one notes is the residual breccia fragments. Thus there is an overabundance of large rock fragments at large distances from a mud diapir eruption site which has little to do with the eruptive breccia thrown into the air (or water in the submarine case). Attempts have been made over the years to correct for this sort of masking phenomenon, with the result that current estimates would generally record an initial mass-distribution proportional to $\exp(-m/m_L)$ where the scaling mass, m_L, is approximately 250 gm (corresponding to an equivalent spherical size of about 5 cm at a rock density of 2gcm^{-3}).

2. Initial Ejection Speeds

Initial ejection speeds of breccia are closely tied to mud and flame ejections, because all three often occur at each eruption of a mud volcano. Ejection speeds of mud are variously estimated. From measured mud flow volumes (V) after an eruption over a time, τ, of a few hours ($2 \times 10^6 \text{m}^3$ of mud in 3 hrs.), and estimating a continuous steady-state emission over the 3 hr. period (which is, in fact, incorrect because one has spurts of emission followed by long quiescent periods before another spurt) a minimum mud ejection speed can be estimated if one also measures the mud crater size (~20-50 m radius) and also assumes that mud emission occurs across the total crater diameter, 2R, (again an incorrect statement-emission is usually over a smaller radius than the total available).

Then the minimum speed is

$$\upsilon_{min} \geq \left(\frac{V}{\tau \pi R^2} \right) \approx 500 m/hr.$$

An improved estimate can be made by noting that the mud column heights (h) during an eruption typically reach 50 m or so, then

$$\upsilon_{min} \approx (2gh)^{1/2} \approx 30 m/sec.$$

In turn, a further refinement can be made on this estimate by noting that ejection most often takes place from restricted areas within a crater, often no more than a quarter of the total area is involved in ejection. Then

$$\upsilon_{min} \approx 4 \times 30 m/sec \approx 120 m/sec.$$

If this estimate of speed occurs at 45° to the horizontal, then the maximum distance a rock fragment can be thrown laterally is around four times the parabolic height, yielding a rough estimate of lateral distance of 500 m. Observations indicate rock fragments out to around 1-2 km from emission centers (but the ambiguity caused by paleo-mud flow transport of fragments and later rain-washing away of mud, leaving only the fragments, does cause some interpretation problems). But, in a general sense, it would seem that typically rock speeds at emission are encompassed by 30-150 m/sec = V_{av}.

⁹Overpressure and Lateral Stress in Sedimentary Formations

One of the substantial hazards occurring during the drilling process is caused by overpressure and/or horizontal stress. Both of these factors can damage and even destroy the well. Therefore an estimation of the probable formation pressure is of great importance. The estimations are especially important in exploration while drilling "wildcats", when one does not have any information from neighboring wells. In this case, in order to get some assessment on the relevant parameters, we have used the results of basin modelling. 2-D models of the South Caspian Basin development have been constructed to estimate overpressure along two crossing profiles and passing across some structures which are the focus of interests of oil companies. At the same time 1-D modelling has been done with the purpose of contouring overpressure for different types of conditions in an area.

To estimate the lateral stress present in the sedimentary pile, use is made of a basement flexural subsidence model combined with inverse methods.

A. Overpressure Hazards

The numerical models used in this work (DTIT and GEOPETII) are based on fluid flow models, compaction and, respectively, changes in porosity, permeability, pressure and so in fluid flow rates with time. A model allows division of the geohistory time into relatively small time-steps, and so one calculates the dynamical processes with time. The calculations can be stopped at any time-step and then be continued with changed input or assumption parameters. For instance, in the 2-D modelling 0.2 MM years time-steps were used until Pliocene time and then, when more precise calculations were needed, the time-step was changed to 0.1 MM years. The fluid pressure is a complex parameter depending mostly on the geohistory, but also on thermal history and on hydrocarbon generation (which creates additional pressure), etc. Those factors are also included in the model. An inverse method is included in the model in order to calibrate the results based on measurements. The model is not described here nor are all the results of modelling presented concerning the development of reservoir characteristics, temperature, organic matter maturity, and oil/gas generation and accumulations. The detailed results are in two previous books (Lerche et al., 1996; Lerche et al., 1997) and in papers (Tagiyev

251

252

et al., 1996; Bagirov et al., 1997). The input data used in the model correspond to the geological description of Chapter 1 and the gas hazards data of Chapter 7.

1. Results of modelling along W-E 12-second cross-section

The section was restored from the 12-second two-way travel time section and crosses almost all of the Caspian Sea from west (Baku Archipelago) to the Turkmenian slope, crossing the Godin uplift.

a. Present-day Fluid Pressures

From the results of the 2-D model the modern hydrostatic pressure increases linearly from 0 at the sediment surface to ~3000 kg/cm^2 (KSC) at a depth of 30 km. The modern lithostatic (overburden) pressures increase linearly from 0 at the surface to ~7000 KSC at a depth of 30 km.

For the Azerbaijan slope zone the fluid pressure increases gradually to a depth of ~8000 m and reaches 1200 KSC at the Maikop (Oligocene-Lower Miocene) surface. The pressure then rises rapidly to 2400 KSC at the base of Maikop layer at a depth of about 12,000 m (Figs. 9.1, 9.2a). The fluid pressure then continues to increase with increasing depth and reaches a maximum value of about 5000 KSC at 24,000 m on the boundary between Upper Jurassic and Lower Cretaceous layers. At greater depths the values return to hydrostatic pressure below ~27,700 m due to a basal sand layer which is input to satisfy the "no-flow" condition across the impermeable basement.

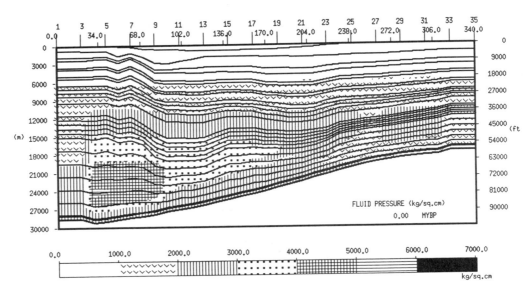

Figure. 9.1 Present-day fluid pressure across the W-E cross-section.

For the abyssal zone the fluid pressure increases gradually to a depth of 11,000 m and reaches 2000 KSC at the Maikop (Oligocene-Lower Miocene) surface. The fluid pressure values then rapidly rises to 3000 KSC at the base of the Maikop layer at a depth of about 15,500 m (Figs. 9.1, 9.2b). At greater depths the fluid pressure continues to increase, reaching a maximum value of about 3500 KSC at about 20,000 m in the Upper Cretaceous layer. Below 24,000 m, the fluid pressure returns to the hydrostatic pressure due to the basal sand layer presence.

For the Godin uplift zone the fluid pressure increases gradually to a depth of 10,500 m and reaches 1700 KSC at the Maikop (Oligocene-Lower Miocene) surface. The fluid pressure then gradually increases to a maximum value of 2000 KSC in the Maikop-Eocene layers at a depth of 12,000-13,000 m (Figs. 9.1,9.2c). As is visible on figure 9.1, fluid pressure has a relative minimum above the Godin volcanic uplift zone. Below 16,000 m, the fluid pressure again takes on hydrostatic pressure values (Figs. 9.1, 9.2c) due to the basal sand presence.

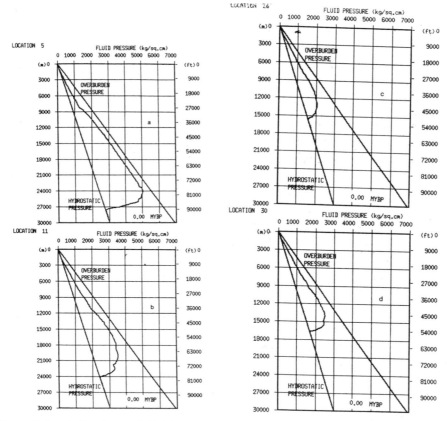

Figure 9.2. 1-D plots of fluid pressure at different locations along the 2-D section:(a) Azerbaijan shelf and continental slope; (b) Central deep-water "abyssal" part; (c) Godin volcanic uplift ; (d)Turkmenistan shelf.

254

For the Turkmenian shelf zone the fluid pressure increases gradually to a depth of 9,200 m and reaches 1600 KSC at the Maikop (Oligocene-Lower Miocene) surface. The fluid pressure then increases rapidly to 2200 KSC at the base of the Maikop layer at a depth of 12,000 m (Fig. 9.1, 9.2a). At greater depths the fluid pressure continues to grow reaching a maximum of 2,450 KSC at 14,000-16,000 m, the depth range of the Upper Cretaceous and Paleocene layers. Below 16,500 m, the fluid pressure returns to the hydrostatic value due to the basal sand layer presence.

b. Present-day Excess Pressure

The picture of excess pressure repeats qualitatively the picture of total fluid pressure. For the Azerbaijan slope zone the excess pressure increases gradually to a depth of 8000 m and reaches 400 KSC at the Maikop (Oligocene-Lower Miocene) surface. The value increases rapidly to 1200 KSC at the base of the Maikop layer at a depth of about 12,000 m (Figs. 9.3, 9.4a). At greater depths, the excess pressure continues to increase and reaches a maximum of 2700 KSC at about 24,000 m on the boundary between the Upper Jurassic and Low Cretaceous layers prior to the decrease to zero at a depth of about 27,700 m due to the basal sand layer.

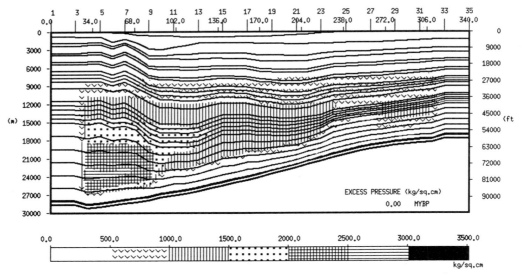

Figure 9.3. Present-day excess pressure across the W-E cross-section.

For the abyssal zone excess fluid pressure increases gradually to a depth of about 11,000 m and reaches 800 KSC at the Maikop (Oligocene-Lower Miocene) surface. The excess pressure then rapidly increases to 1600 KSC at the base of the Maikop layer at a depth of about 15,500 m (Figs. 9.3, 9.4b). At greater depths the excess pressure continues to increase, reaching a maximum of about 1600 KSC at 18,000 m in the Upper Cretaceous layer, before decreasing to zero at 24,000 m due to the presence of the basal sand layer.

For the Godin uplift zone the excess pressure increases gradually to a depth of 10,500 m and reaches 700 KSC at the Maikop (Oligocene-Lower Miocene) surface, before increasing rapidly to 1000 KSC at the base of Maikop layer at a depth of about 12,000 (Figs. 9.3, 9.4d). At greater depths excess fluid pressure continues to increase, reaching a maximum of about 1150 KSC at 13,000-14,000 m, the Upper Cretaceous and Paleocene layers, before decreasing to zero at 16,500 m depth due to the presence of the basal sand layer.

2. Results of N-S cross-section modelling

This profile crosses the 12-second cross-section in the south-west part near the Sabail structure, and then goes to the north-east direction, passing the Sabail, Umid, Shakh-Deniz, Haji-Zeinalabdin and Oguz structures. The geological data taken from the 6-second seismic cross-section, of 150 km length, were used as a basis for modelling. The cross-section reflects the SCB geological structure to a depth of about 15 km. Missing information was obtained by interpolation (Lerche et al., 1997, (Ch. 3); Bagirov et al., 1997). The results of the modelling concerning fluid pressure are very similar to those for the west part of the W-E profile.

Two faults are observed on the cross-section. Therefore two extreme cases were run: when the faults are closed, and when the faults have high permeability.

a. Present-day Fluid Pressures

From the results of the 2-D model the modern hydrostatic pressure increases linearly from zero at the sediment surface to ~3000 kg/cm² (KSC) at a depth of 30 km. For comparison, the modern lithostatic (overburden) pressures increase linearly from zero at the surface to ~7000 KSC at a depth of 30 km.

Comparing figs. 9.5a and 9.5b, corresponding to the cases of closed and open faults respectively, one can observe a negligible difference between fluid pressure patterns closest to the fault zones. In general the fluid pressure increases gradually to 1300-1500 KSC at the depth range 8,000-9,500 m of the Maikop surface (Oligocene-Lower Miocene). At the Oguz structure, the Maikop surface corresponds approximately to a depth of 7000 m, and fluid pressure there is about 900-1000 KSC (Figs. 9.5a-b, 9.6a-e). At greater depths the pressure gradient increases and the absolute pressure rises rapidly.

The fluid pressure continues to increase with increasing depth and reaches a maximum on the boundary between Upper Jurassic and Lower Cretaceous layers. In the accretion zone this boundary corresponds to depths of about 23,000-25,000 m where fluid pressure reaches 4,500-5,500 KSC, while in the abyssal zone the boundary occurs at 19,000-21,000 m with the fluid pressure at about 3,500 KSC.

256

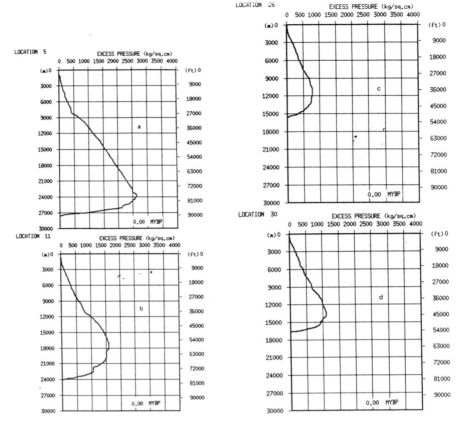

Figure 9.4. 1-D plots of excess pressure at different locations along the W-E 2-D section: (a) Azerbaijan shelf and continental slope; (b) Central deep-water "abyssal" part; (c) Godin volcanic uplift ; (d)Turkmenistan shelf.

b. Present-day Excess Pressure

The picture of excess pressure qualitatively repeats the picture of total fluid pressure. In general, the excess pressure increases gradually to a depth of 8,000-9,500 m and reaches 400-600 KSC at the Maikop (Oligocene-Lower Miocene) surface (Figs. 9.7a,b, 9.8a-d). For the Oguz structure this transition point occurs at a depth of 6 km, where excess pressure reaches the value 250 KSC (Fig. 9.8e). The excess pressure then increases rapidly and reaches a maximum on the boundary between Upper Jurassic and Lower Cretaceous layers, prior to decreasing to zero at the basal sand layer. In the accretionary zones, the maximum (2400-2800 KSC) corresponds to depths of 23-27 km; in the abyssal zone the maximal value is 1600-1700 KSC at depths of 18-20 km. Simultaneously, a significant difference can be observed between the excess pressure patterns near the faulting zones, when the permeability of the faults is changed. Around open faults no excess pressure is observed (Fig. 9.7b), while at a depth of 9 km an excess pressure of about 500 KSC is observed at the locations of closed faults.

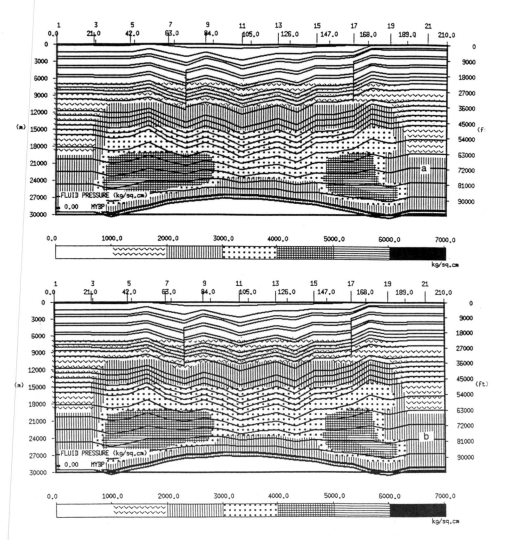

Figure 9.5. Present-day fluid pressure across the N-S 2-D section: a) faults closed; b) faults open with high permeability.

3. Results of 1-D Modelling

The study area is a north-western portion of the South Caspian sedimentary basin. Geographically the region is enclosed between latitudes 48°N and 51°N, and longitudes 36°E and 42°E (Fig. 9.9). Specific conditions are determined by a number of geological parameters such as low heat flow (20-50 mWm^{-2}), high sedimentation rate (to 1.3 km/Ma), great thickness of sedimentary cover (20-30

258

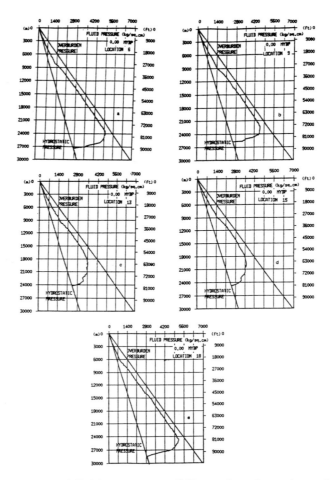

Figure 9.6. 1-D plots of fluid pressure at different locations along the N-S 2-D
section: a) Sabail; b) Umid; c) Shakh-Deniz; d) Hajy Zeinalabdin;
e) Oguz.

km), and thick argillaceous lithologies constituting up to 90% of the section
uncovered by drilling.

The predominance of argillaceous shale and clay material throughout
the study area, and estimated to occur as early as the Jurassic, argues for the
presence of fluid overpressuring. The competition is between the rate of
sedimentation, the low permeability of shales which generally impedes fluid
flow, and the rise in overpressure which, if too large, will lead to fracturing of
sediments with a concomitant loss of overpressure to below the fracture limit.
Most types of sedimentary rocks fracture when the total fluid pressure is a
fraction of between about 0.75-0.85 of the total overburden (Jaeger and Cook,

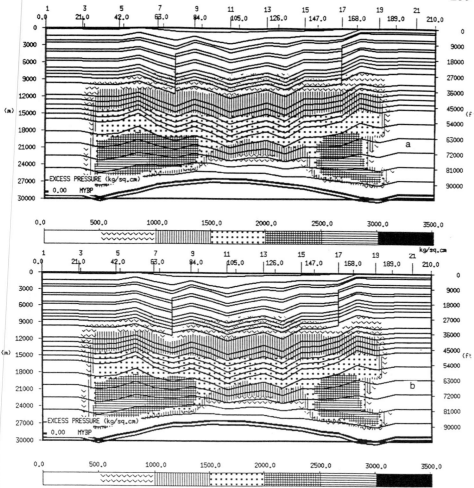

Figure 9.7. Present-day excess pressure across the N-S 2-D section: a) faults closed; b) faults open with high permeability.

1976) - although extreme situations are known (Bour et al., 1995) where the lateral stress exceeds the overburden load, so that the principal rock failure direction is no longer effectively vertical. In such cases fluid pressure has reached 0.95 of total overburden load.

Figure 9.10 provides in situ pore pressure measurements with depth for the three wells Bulla-Deniz, Bahar-Deniz, and Neft Dashlary, indicating a lowering of fluid pressure to the north of the study area from about 1000 atm at 5.5 km depth at the Bulla-Deniz well, to about 620 atm at the same depth at Neft Dashlary. The vertical pressure gradient is then from about 0.18 atm/m ($\cong$ 0.9 psi/ft) (Bulla-Deniz) to about 0.11 atm/m (0.55 psi/ft) (Neft Dashlary). The total distance between Bulla-Deniz and Neft Dashlary is about 100 km, so the corresponding horizontal fluid pressure gradient is around 0.35 atm/km directed

260

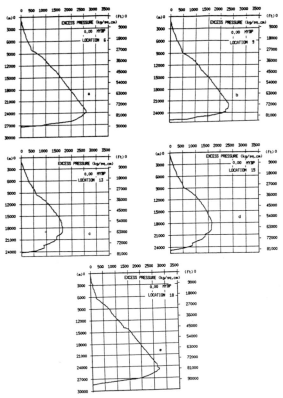

Figure 9.8. 1-D plots of excess pressure at different locations along the N-S 2-
D section: a) Sabail; b) Umid; c) Shakh-Deniz; d) Hajy
Zeinalabdin; e) Oguz.

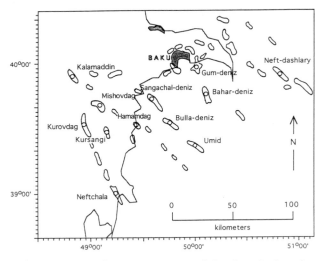

Figure 9.9. Study area - north-western part of the South Caspian sedimentary
basin with the 12 control wells marked, together with the outlines of
hydrocarbon fields.

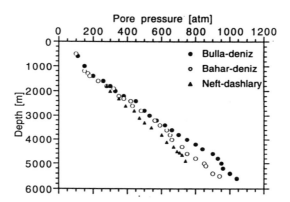

Figure 9.10. Pore pressure vs. depth in wells Bulla-deniz, Bahar-deniz and
Neft-Dashlary (from Abasov et al., 1991).

from the high pressure region around Bulla-Deniz towards lower pressure in the
northeastern sector of the study area. Across the study area as a whole, the total
vertical fluid pressure gradient for the twelve wells has an average of 0.8 psi/ft.

There is a generally accepted thought that the rise of overpressure
across the basin, together with the massive gas generation, are the two likely
main causes for the presence of the observed mud volcanoes, of which more
than 220 are recorded in the SCB.

Figure 9.11 illustrates the excess fluid pressure build-up with time at
well Bulla-Deniz. The rapid deposition of the Jurassic shales (9 km in about 50
Ma) causes an early build-up of excess pressure to about 250 kgcm^{-2} (1 kgcm^{-2}
$\cong$ 1.03 atm) at 150 MYBP. However, the lack of any significant further
deposition during the Cretaceous gives plenty of time for this overpressure
build-up to "bleed off", reducing undercompaction to more nearly normal
behaviors. In the Tertiary, the steady deposition of a further 4 km of
argillaceous shales from 65 MYBP to 3.4 MYBP causes a secondary rise in
overpressure to a maximum of about 170 kgcm^{-2} between an overpressure top
at 6 km to a basal depth of 14 km. The massive shale deposition from mid-
Pliocene time onward exacerbates the rise of overpressure, yielding a
continuously increasing overpressure with time which rises to as high as around
800 $\pm$ 50 kgcm^{-2} in the late Pliocene to early Quaternary through to the present-
day. The total fluid pressure rises nearly to the fracture point ($\sim$0.85) for

typical consolidated, brittle, failure but, given the ductile nature of shales, the deformation of sediments and associated mud volcanoes would seem to be the prevalent path by which massive rock fracture failure is avoided. For each of the 12 wells in the study area, similar developments of overpressure evolution with time have been evaluated. In this way contours can be given of excess pressure development with time across the basin at different fixed depths or, alternatively, depth values to a given isobaric level can be drawn, thereby indicating the relative equivalent hydraulic head driving fluid flow across the basin. To provide a basis for evaluating the evolution of the fluid pressure development across the basin in relation to sediment fill, relative to a structural map at base of sedimentary cover (Figure 9.12a), three major groupings of sediment isopachs are looked at here (Figures 9.12b,c,d). First, at the end of the dominant Jurassic deposition, the western, southern and central regions had accumulated about 7-7.5 km of sediments by shedding from the Talysh-Vandam high region, with the Apsheron region receiving less than about 5-6 km, leading to a local high in the Bulla-Deniz region of order 2.5 km relative to sediment supplied to Baku (Figure 9.12b).

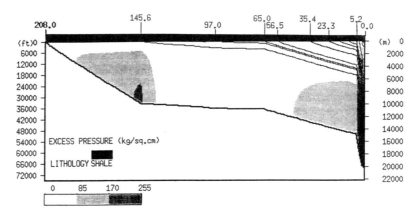

Figure 9.11. Excess fluid pressure build-up through time for well Bulla-deniz.

By late Neogene time, the combined isopach thickness across the section from Cretaceous through Neogene amounted to some 5 km in the region of Bulla-deniz, and about 8-9 km along the Greater Caucasus-Apsheron peninsula region, so that the relative high of 2.5 km at Bulla-Deniz is now reversed to a depression of about 1.5-2.5 km relative to the structural high which follows parallel to sub-parallel to the orogenic development of the Caucasus (Figure 9.12c).

The massive deposition in the mid-Pliocene through Quaternary resulted in a isopach thickness of 5-7 km in the southeastern portion of the area (towards the Turkmenian border) with a thinner section of around 4-5 km across much of the north, parallel to the Greater Caucasus. The depression in the central part of the basin is then filled by the extra 1-3 km of late Pliocene/Quaternary sediments, with local variations due to transpressive thrust of the Talysh-Vandam wedge under the Greater Caucasus plate after Tethys Sea closure in late Miocene/early Pliocene (Figure 9.12d).

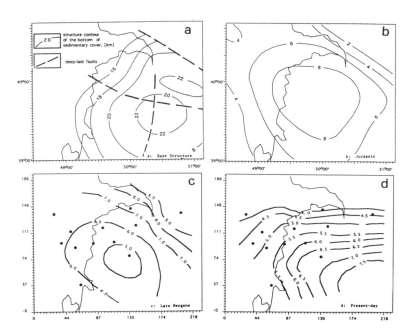

Figure 9.12. (a) Structural map at base of sedimentary pile (from Gasanov et al., 1988); (b) Schematic isopach map of Jurassic deposits (from Bagir-zadeh et al., 1988); (c) Schematic isopach map of Cretaceous-Paleogene-Miocene interval of the sedimentary cover; (d) Schematic isopach map of Pliocene-Quaternary interval of the sedimentary cover. Contours are in km.

a. Fixed depths, variable excess pressure with time

The development of the fluid pressure in the region over geological time can be found in Tagiyev et al., (1997) and Lerche et al. (1996). Here we show only present-day variations of the fluid pressure (in units of atmospheres) at depths 6, 8 and 10 km respectively. There is approximately 40 atm increase between 6 and 10 km, but the pressure values are typically scattered in the range 80-140 atm.

The patterns of overpressure for those depths are shown in figure 9.13, caused by massive deposition of predominantly shaly sediments. The highest overpressure is observed in the central part around the Bulla-Deniz area. At a depth of 6 km the overpressure reaches 500 atm, and increases with increasing depth. The overpressure decreases to the north and, in the area of Apsheron-Balkhan uplift, drops to 350-400 atmospheres at 6 km depth. These values correspond to the results of 2-D modelling and are useful in order to show the lateral changes of overpressure at different depths. As well as the horizontal gradient of excess pressure, the vertical gradient is now increased to around 50 atm/km between 6 to 10 km depth (i.e. around 0.25-0.3 psi/ft of <u>excess</u> fluid pressure) which, when added to the hydrostatic pressure gradient of about 0.5 psi/ft., yields an estimated vertical fluid pressure gradient of 0.75-0.8 psi/ft - as is generally observed in the study area.

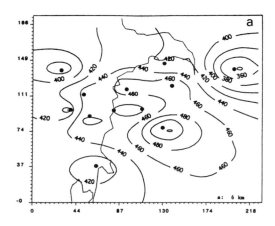

Figure 9.13. Contour maps of excess fluid pressure at present-day at depths: (a) 6 km.

b. Isobaric pressures, variable depths with time

An appreciation of which formations are most likely to be responsive (in terms of fluid flow) to applied excess pressure can be attained by constructing depth contours with time across the basin at which a fixed isobaric total fluid pressure is reached. A constant depth across the study area would indicate no lateral pressure gradient. Thus the difference in elevations across the basin in order to reach a prescribed isobaric value gives an idea of the differential hydraulic head driving fluid migration.

Shown in figures 9.14a-b, are the development of contours of depth elevation to reach total isobaric pressures of 900 atm and 1200 atm, respectively, at present-day.

By present-day, as shown on figures 9.14a (900 atm isobar) and 9.14b (1200 atm isobar), the driving head is reduced to 0.2-0.6 km to the northeast from the central part of the basin at 900 atm, and to about 1 km head at 1200

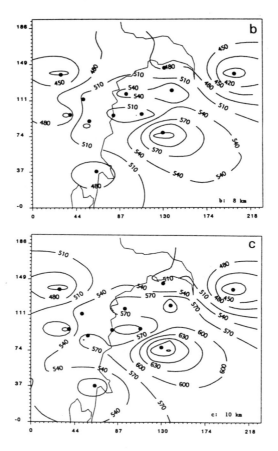

Figure 9.13. Contour maps of excess fluid pressure at present-day at depths:
(b) 8 km; (c) 10 km. Contours are in atmospheres.

atm; with northwesterly and southwesterly head drives of around 0.4-0.6 km at the 1200 atm isobar value, and 0-0.4 km head at the 900 atm isobar value.

Buryakovsky and Djevanshir (1983) have compiled average sand/shale ratios for wells in the study area and grouped them into three sub-areas, called Areas I, II and III respectively, as shown on figure 9.15. To the typical well depth of around 5-6 km, the average sand/shale ratio is around 50:50 in Area 1, around 30:70 in Area II, and around 10:90 in Area III. These domains are precisely those with the highest overpressure today (Area III, as typified by Bulla-Deniz), medium overpressure today (Area II, as typified by Bahar-Deniz), and low overpressure today (Area I, as typified by Neft Dashlary), and are in accord with the estimated directions of lateral fluid pressure gradients since Pliocene time.

In addition, data on shale porosity variations with depths have been compiled in the wells by Bredehoeft et al. (1988) and, as depicted in Table 9.1 for the average values from wells in each of the sub-areas, there is a systematic tendency for shale porosity to be retained to a higher degree in Area III than in

Area II, and retained more in Area II than in Area I, again indicative of the direction of recent fluid flow from the central part of the basin to the northeast.

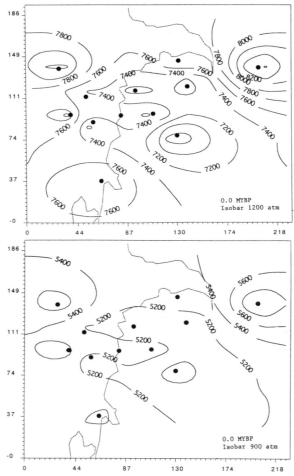

Figure 9.14. Isobaric depth contour maps across the area for 900atm and 1200atm total fluid pressure values at 23.3MYBP, 5.2MYBP, 3.4MYBP, 1.8MYBP and present-day, respectively. Contours are in meters.

B. Horizontal Stress

Another complication which can occur during drilling is the horizontal stress. Some horizontal stress can be associated with high overpressure in the zones of intensive folding or faulting (Bour et al., 1995). But that correspondence should be investigated separately for each special case. Another reason for high horizontal stress is rising mud diapirs, already considered in Chapter 3. As was shown in the example of the Abikh diapir, the values of the accumulated lateral stresses in the surrounding sediments are up to 4.10^{10} Pa due to the rise of the diapir of about 10 km diameter.

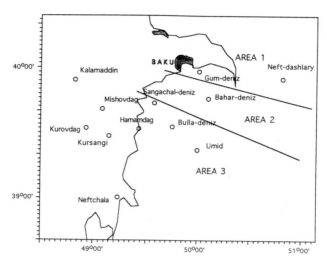

Figure 9.15. Three sub-areas delimited by various average sand/shale ratios and porosity profiles (from Bredehoeft et al., 1988).

Similar constructions have been done for the Vezirov diapir (Lerche et al., 1996). As is shown in fig. 9.16 accumulated stress in the deep zones can again reach about 4.10^{10} Pa. In the shallower zone reached by the well, this value is less and the stress usually spreads over 2-3 diapir diameters. However, an accurate estimation should be done for each zone, adjusted to each mud diapir. The major <u>regional</u> stress is caused by the tectonic setting of the area. As described in Chapter 1, the South Caspian Basin is due to closure of the Tethys Ocean. Since Oligocene time, and especially during Neogene, the Kurian-South Caspian zone of the crust was subject to the influence of orogenic processes which developed in the Greater and Lesser Caucasus and in Talysh. As was shown in Chapter 1, two main directions of pressure took place: (i) Western - in the direction to the West Caspian deep-laid fault (or in the direction of the Talysh-Vandam buried uplift zone); and (ii) Sub-northern - in the direction to the Greater Caucasus - Apsheron-Balkhan prolongation.

The qualitative description given above of the processes causing the unique properties of the SCB demands both quantitative control and modeling of the processes. To achieve these aims, and to provide a quantitative estimation of parameters involved in the flexural plate motion, the processes were modeled using a flexural plate code (FLEX4, developed at the University of South Carolina).

Models of compensation by flexure are based on the assumption that the lithosphere has some finite strength. Loads are supported by the mechanical strength of the lithosphere as well as by buoyant forces (Turcotte, 1979).

268

Table 9.1. Shale porosity vs. depth for Areas I, II and III (from Bredehoeft et al., 1988)

Depth (m)	Porosity (%)		
	Area I	Area II	Area III
750		27.0	29.0
	31.0		
1,250		21.5	23.5
	26.0		
1,750		16.0	19.0
	22.0		
2,250		12.0	14.0
	20.0		
2,750		8.5	13.0
	18.5		
3,250		7.0	12.0
	16.5		
3,750		5.5	10.4
	16.0		
4,250		4.5	9.5
	16.0		
4,750		3.0	8.5
	13.5		
5,250		-	-
	12.3		

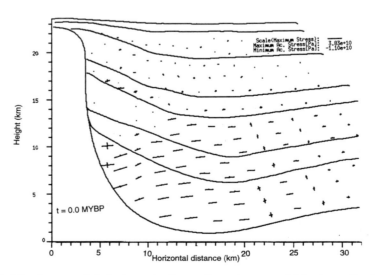

Figure 9.16. Accumulated stress in the sediments around Vezirov diapir.

Linear elastic models, based on Hooke's law, are such that strain is directly proportional to stresses (Turcotte and Schubert, 1982); deformation occurs instantaneously when a given stress is applied and there is complete recovery when the stress is removed (no permanent deformation).

The model assume that the lithosphere is an infinite plate. Such a lithosphere must be horizontal prior to loading, so that the plate cannot have an initial dip angle or initial bending moment. Only two parameters then control the bending of an infinite elastic plate: the flexural rigidity (D), and the load (V), a situation which limits the number of plate geometries that can be produced. A lithospheric plate of <u>finite</u> length has four more parameters entering the elastic model: plate length (L), initial dip angle (θ), initial bending moment (M), and lateral stress (P) (Fig. 9.17).

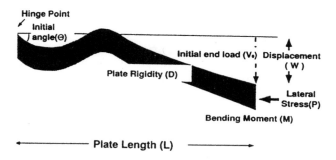

Figure 9.17. A flexural model with compressive thrusting.

In a two-dimensional situation, one end of a plate is taken as a hinge point, normally representing a geological boundary. Compressive or extensional stress is applied at the other end of the plate. The hinge point position is either specified based on geological requirements (e.g. a bounding fault), or can be adjusted vertically and laterally (see later) when the hinge point position is then a parameter to be determined. The sediment cover is applied on the plate as a vertical force. The deformation of the plate due to the applied stresses is calculated relative to the hinge point as zero displacement (Fig. 9.17). The unloaded plate configuration is an indication of the basement geometry that existed prior to any sediment loading or thrust. The allowed variation in pre-loaded basement shapes can produce greater variability in the final (loaded) basement geometry than can an initially flat (infinite) plate.

The detailed mathematical analysis of the elastic flexural plate equations is provided in Appendix A.

The flexural response of the basement to sediment loading and to lateral stress due to compressive thrusting or extensional rifting, depends on the basic parameters: rigidity of the basement D, initial bending moment M_0, initial end-point load V_0, dip angle of the basement at the hinge point θ, length of the basement L, lateral stress P applied to the end of the basement, and the density difference between mantle and sediment $\Delta\rho$. The deformational process is taken to be independent of depositional rate; i.e. the plate response is sufficiently rapid that at each instant of time the steady-state response is an accurate representation. Therefore, all of the parameters in the model are taken as constants through geological time. However, the parameters in the model are related to spatial position along the basement. The parameters divide into three groups: the first sets the overall scales of the basic parameters; the second determines whether the basic parameters systematically increase or decrease along the basement; the third is used to simulate a sinuous shape of basement deformation. Simulations then allow for the effects of regional geological events on basement and also of the effects of local events on basement deformation, such as local intrusions of volcanic material, metamorphic phase changes, etc.

Inverse methods determine the parameters in models (Cogan et al., 1989; Tang et al., 1990; Tang et al., 1992; Tang and Lerche, 1992). Parameter values are adjusted in a systematic non-linear optimization procedure to provide a close approximation to the observed, loaded basement profile determined by the best fit between calculated and observed basement profiles. Tang and Lerche (1995) provide a detailed description of the inverse method and a précis is given in Appendix A. With the best parameters so determined, the inverse method then removes sediment cover to determine an unloaded basement profile.

To investigate the effects of individual sedimentary layers on basement deformation, the basement evolution with geological time is modeled as follows: first, the inverse model is used to determine the flexure parameters which best fit the observations; second, by running a forward model with the best values of the parameters (without sediment cover on the basement) one obtains an unloaded profile; third, by adding sedimentary layers to the unloaded profile in a sequential fashion, the basement is allowed to "bend" and the basement shape modified by sedimentation. When all sedimentary units are laid down, the final calculated basement shape will be that having the minimum mismatch to the observed profile. As the load of each layer is added, modification of the basement shape occurs, yielding the deformation due to the change in differential sediment loading over a given geological time interval.

Here we treat only with an elastic plate model with plate parameters held constant along the plate. If such an elastic model can describe the present-day basement configuration there is no need to invoke more complicated models. The point here is that the insensitivity of deformation to the details of rheology permits the highly successful use of elastic plate models.

The simplest situation, which illustrates the logic of the procedure, is the case of an elastic plate whose rigidity is constant (termed elastic model 1). Flexure of a finite elastic beam of length L, anchored at one end, due to the force of a single applied end-load is dictated by six parameters: flexural rigidity, D; bending moment, M; dip angle of the anchored end, θ; end load, V; length, L; and lateral stress, P (Appendix A, and figure 9.17). Initial application of this model is limited by the assumption that the total load of the basin can be approximated by a point load applied at the end of the plate. The single load model has to be modified to account for differential lateral loading (Cogan et al., 1989).

Once a basement horizon is picked (usually on the basis of seismic and gravity data) and the hinge point chosen (often taken to be the highest point of the basement on the section), the sediment load overlying the basement is then divided into a series of blocks of equal width (figure 9.18). The lines between the blocks are drawn from the seismic datum to the top of the basement. If the porosity is not measured then we assume as a default position that the porosity decreases exponentially with depth:

$$\phi(z) = \phi_0 \exp(-z/l)$$

where $\phi(z)$ is the porosity at a given depth z, ϕ_0 is the porosity at the depositional surface and l is a scaling length (Sclater and Christie, 1980). The values of the surface porosity and scaling length have to be specified. For default values we use Sclater and Christie's suggested values of $\phi_0 = 0.56$ and l = 2300 m, for sandy-shales. The total sediment load within each block is then calculated as follows:

$$V_S = gA \int_0^z [\rho_w \phi(z) + \rho_s(1 - \phi(z))]\, dz \qquad (9.1)$$

where V_S is the total sediment load within the block, g is the acceleration due to gravity, A is the cross-sectional area of the block, ρ_s is the density of the sediment matrix material, and ρ_w is the density of water. If a basin is partially filled by water, we have

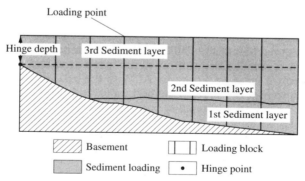

Figure 9.18. Sketch of point loading relative to hinge point.

to consider the effect of water loading on deformation of basement. The weight of water of column height h_w in the basin is calculated from:

$$V_W = gA\ \rho_w h_w \qquad (9.2)$$

The total load on the basement within each block is then:

$$V_T = V_s + V_w \qquad (9.3)$$

A theoretical basement profile is calculated for the observed loading scheme with the parameter values set at some initial values. The theoretical profile is compared to the observed profile using a standard least-squares fit. The average degree of mismatch is represented by the MSR, or mean square residual. The lower the MSR, the better the fit. (The unit of MSR is length squared, but we take the logarithm of MSR in order to simplify the calculations. Henceforth, all MSR values refer to the logarithm of a number whose unit is meters squared). After an MSR is calculated for the initial parameter estimates, an inverse method (Appendix A) is employed to adjust all the parameter values simultaneously to minimize the MSR. The length, L, of the plate is treated differently in the inverse approach than in the forward model. It is possible that the plate extends beyond the limits of the data, but cannot be shorter than the observed end of the plate. Therefore, plate length can be varied, giving rise to an "added length" which refers to the segment of plate added onto the end of the observed plate furthest from the hinge. Likewise the initial choice of hinge point position may be mistaken, accordingly the hinge point can be adjusted both vertically and laterally using the non-linear procedure to minimize the MSR.

When the best values of the parameters have been determined the model is run for a final time. The theoretical basement profile is determined, the final MSR calculated, and the pre-loading basement profile obtained (Cogan et al., 1989; Tang et al., 1990).

In this chapter we omit the details of the modeling process and the results for the basin subsidence for the South Caspian Basin done for the 12-second seismic section. Those results can be found in Lerche et al. (1996, Ch. 1) and Nadirov et al. (1997). Here we are interested only in the horizontal stress present in the sedimentary pile due to plate motion. Figure 9.19b indicates a well-defined minimum in MSR at log lateral stress (N/m) of 12.3±0.3. This lateral stress along the line of section is the result of the superposition of two stresses: one acting from the west (Talysh-Vandam), the other from the north (Apsheron-Balkhan).

Referring to the sketch of figure (9.20), with a lateral stress of P_{AB} (P_{TV}) at right angles to the Apsheron-Balkhan region of the Greater Caucasus (the Talysh-Vandam region of the deep-laid-fault), the components in the line of section add to yield

$$P = P_{AB}\cos\theta_2 + P_{TV}\cos\theta_1 \qquad (9.4)$$

while the components perpendicular to the line of section approximately balance so that little strike-slip occurs i.e.

$$P_{AB} \sin\theta_2 = P_{TV} \sin\theta_1 \qquad (9.5)$$

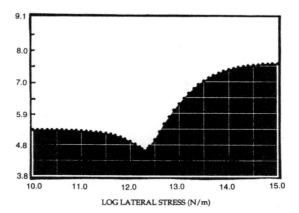

LOG LATERAL STRESS (N/m)

Figure 9.19. Sensitivity analysis plots of MSR (ordinate values) versus lateral stress.

Using equation (9.5) in equation (9.4) permits a determination of the magnitudes of both the stresses produced by the Talysh-Vandam and Apsheron-Balkhan tectonic events in terms of the stress P ($\cong 10^{12.3}$ N/m) inferred along the plate.
Then

$$P_{TV} = P \sin\theta_2/\sin(\theta_1 + \theta_2) \qquad (9.6a)$$

and

$$P_{AB} = P \sin\theta_1/\sin(\theta_1 + \theta_2) \qquad (9.6b)$$

Direct measurement from figure 9.20 yields $\theta_1 \approx 33°$; while $\theta_2 \approx 100°$. Hence: $P_{TV} = 0.25P$, $P_{AB} = 0.71P$. Accordingly, about 3/4 of the thrust on the line of section arises from the Apsheron-Balkhan compression, with the remaining 25% due to the western thrust from the Talysh-Vandam area.
In terms of a critical lateral stress, P_c, given by (Turcotte and Schubert, 1982)

$$P_c = 2[gD(\rho_m - \rho_o)]^{1/2} \cong 10^{14} \text{ N/m} \qquad (9.7)$$

the actual stress in the line of section is only about 3% of P_c. Thus there is not enough applied stress to cause a critical forebulge or uplift of the basement, but there is enough to make the influence of the stress felt in the flexure of the plate, which is otherwise dominated by rigidity, bending moment and dip angle in relation to the applied sedimentary load.

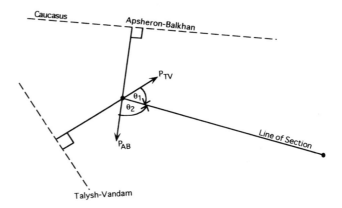

Figure 9.20. Sketch of lateral stress acting perpendicular to the Apsheron-Balkhan region and to the Talysh-Vandam region, and resolved into a compressive thrust in the line of section.

Those constructions have been done for the specific location of the seismic line. In order to obtain more accurate assessments of the lateral stress one has to make similar models for any set of sections crossing an area of planned drilling.

C. Conclusions

The region of the South Caspian Basin is very complex for drilling. Present-day overpressure is high. All models and investigations described above have a large-scale, regional character. The purpose here was to show the overall character of the overpressure and stress in the region. Overpressure in the Pliocene-Quaternary formations is relatively small. On average the gradient of increment of the excess pressure is about 40-60 atm/km of depth.

Starting from Miocene formations and deeper, excess pressure starts increasing rapidly, with a gradient of about 100-150 atm/km. This increase can be explained by lithological factors. The presence in the Middle Pliocene of laterally extended sandy horizons, which are the reservoir rocks of the Productive Series, released significant overpressure from the formations. While the sedimentary units of the lower formations are mostly shaly and the sandy bodies are mostly lens-shaped, they do not extend laterally very far. Therefore their role as a fluid flow channels in pressure release is negligible.

As for the faults, most of them are closed, as the models showed. Even if the faults are "open" their influence has a local character and does not spread far laterally. So one of the simpler conclusions is that a driller has to expect overpressure in the Productive Series. But drilling into the Miocene is much more complex. The overpressure in the deeper horizons is higher, as the onshore data also testify. For example, tests in the Umbaki-Garadag area showed formation pressures of 550-600 atm at a depth of 3 km.

As for the regional lateral stress, it arises mainly from the Apsheron-Balkhan and Talysh-Vandam zones. The influence of the northern plate is higher (Fig. 9.20). Therefore, the closer to the edge of the basin one goes, the higher the expected lateral stress. Rising diapirs also produce significant lateral stress but of a more local nature. However, for each specific area it is necessary to make assessments of the overpressure and of the horizontal stress before drilling.

Appendix A

1. *The Governing Equations for Elastic Flexure with Constant or Variable Plate Rigidity*

For a two-dimensional elastic plate the vertical displacement, $w(x)$, is controlled by the downward load per unit length, $q(x)dx$, acting between x and $x + dx$ (where x is the horizontal coordinate), by the bending moment $M(x)$, and by the applied horizontal force $P(x)$ through (Turcotte and Schubert, 1982) the equation

$$\frac{d^2 M}{dx^2} = -q(x) + \frac{d}{dx}\left\{P(x)\frac{dw}{dx}\right\} \tag{A1}$$

For a plate of local thickness $h(x)$, with a Young's modulus $E(x)$, and a local Poisson's ratio $v(x)$, the connection between the bending moment $M(x)$ and the displacement $w(x)$ is given through (Turcotte and Schubert, 1982)

$$M(x) = - D(x) \frac{d^2 w}{dx^2} \quad , \tag{A2}$$

where the rigidity $D(x)$ is given by

$$D(x) = E(x)\, h(x)^3 / [12(1 - v(x)^2)] \tag{A3}$$

Substituting equation (A2) into equation (A1) gives the equation governing displacement of the plate in the form

$$\frac{d^2}{dx^2}\left[D\frac{d^2 w}{dx^2}\right] + \frac{d}{dx}\ [\ (P(x)\frac{dw}{dx}\ ]\ - q(x) = 0 \tag{A4}$$

For lithospheric material of density ρ_m, overlain by material of average density ρ_0, Archimedes' Principle provides an expression for the net downward load per unit length of (Turcotte and Schubert, 1982)

$$q(x) = - (\rho_m - \rho_0) \, gw \; , \tag{A5}$$

where g is the acceleration due to gravity.

Equation (A5) when substituted in equation (A4) yields

$$\frac{d^2}{dx^2} \left[D \frac{d^2 w}{dx^2} \right] + \frac{d}{dx} [\, (P(x) \frac{dw}{dx}] \; + (\rho_m - \rho_0) wg = q_1(x) \; , \tag{A6}$$

where $q_1(x)$ is the applied sedimentary load on the plate.

There is, a priori, no reason to suppose that either the rigidity, D, the compressive thrust, P, or the anomalous density contrast, $(\rho_m - \rho_0)$, are independent of spatial position along the plate. For instance for a plate which extends under both oceanic and continental domains the overlying density, ρ_0, could range from that of water $(\rho_0 \approx 1 \; gmcm^{-3})$ to that of continental crust material $(\rho_0 \approx 2.7 - 3 \; gmcm^{-3})$ with position along the plate (Turcotte and Schubert, 1982). Equally, for a plate of variable thickness, but constant Young's modulus, the plate rigidity would vary as the cube of the plate thickness, and so on. The response is required of an elastic plate when the "parameters" describing the plate's physical characteristic behavior (viz. D, P and $\Delta\rho \equiv (\rho_m - \rho_0)$) are location dependent.

An avenue of approach is available. We can attempt numerically to find the general solution to

$$\frac{d^2}{dx^2} \left[D \frac{d^2 w}{dx^2} \right] + \frac{d}{dx} [\, (P(x) \frac{dw}{dx}] \; + g\Delta\rho(x)w = q_1(x) \; , \tag{A7}$$

directly with D, P, and $\Delta\rho$ varying, and then compute the effects of the spatially variable sediment loading.

The boundary conditions on the plate are: At the hinge point (x = 0) the plate displacement is zero, i.e.

$$w = 0, \; \text{on } x = 0; \tag{A8a}$$

and the plate hinge angle is θ, i.e.

$$\frac{dw}{dx} = \tan\theta, \; \text{on } x = 0. \tag{A8b}$$

On the end of the plate, x = L, the conditions are that the bending moment M_L is given, i.e.

$$M_L = - D \frac{d^2w}{dx^2}, \text{ on } x = L; \tag{A8c}$$

and the line load V_o at the end of the plate is given, i.e.

$$\frac{d}{dx}\left(D \frac{d^2w}{dx^2}\right) + P \frac{dw}{dx} = - V_0, \text{ on } x = L. \tag{A8d}$$

The problem of numerical integration is best handled by constructing the coupled implicit equations

$$w(x) = x\tan\theta - \int_0^x (x - x') M(x')/D(x') \, dx' \tag{A9a}$$

and

$$M(x) = M_0 + xP(x=L) \tan\theta + xR$$

$$+ \int_0^x P(x')\left[\int_0^{x'} M(x'')/D(x'') \, dx''\right] dx'$$

$$+ \int_0^x (x-x') [(\rho_m-\rho_0)gw(x') - q_1(x'] \, dx' \tag{A9b}$$

where the constant R is related to the end point load V_0 through

$$R = V_0 - \int_0^L [(\rho_m-\rho_0)gw(x') - q_1(x')] \, dx' \tag{A10}$$

M_0 is the bending moment at the hinge point ($x = 0$), and $P(x = L)$ is the value of $P(x)$ at the end of the plate on $x = L$. The bending moment, M_L, on the end of the plate is related to M_0 through equation (A9b) evaluated on $x = L$, viz.

$$M_L = M_0 + L \tan\theta P(x = L) + LR$$

$$- \int_0^L P(x)\left[\int_0^x M(x')/D(x') \, dx'\right] dx$$

$$+ \int_0^L (L - x) [(\rho_m - \rho_0) gw(x) - q_1(x)] \, dx \qquad (A11)$$

Equations (A9a) are integral representations of equation (A7), automatically satisfy the boundary conditions, and can also be easily integrated numerically once the general functional forms of $D(x)$, $P(x)$ and $\Delta\rho(x)$ ($\equiv \rho_m - \rho_0$) are specified.

To allow for spatial variations, write D, P and $\Delta\rho$ in the forms

$$D(x) = D_0 \exp \{\beta x + b_1 \sin\pi x/L + \beta_2 \sin 2\pi x/L + \ldots\} ,$$
$$P(x) = P_0 \exp \{\gamma x + \gamma_1 \sin\pi x/L + \gamma_2 \sin 2\pi x/L + \ldots\} ,$$
$$\Delta\rho(x) = \Delta\rho_0 \exp \{\delta x + \delta_1 \sin\pi x/L + \delta_2 \sin 2\pi x/L + \ldots\} , \qquad (A12)$$

where D_0, P_0, $\Delta\rho_0$ are scaling constants and β, γ, δ measure the overall trend of change in elastic parameters along the plate, while β_1, β_2 ...., γ_1, γ_2, ..., δ_1, δ_2, ..., measure local scale fluctuations relative to the trend values. The advantage of the forms (A12) is that they keep D, P, and $\Delta\rho$ manifestly positive and, by the addition of oscillatory factors, the degree of resolution of each parameter can be determined as well as allowing for a plate to start rigid on $x = 0$, become less (more) rigid in the middle, and more (less) rigid at its end $x = L$, as the need dictates. In the investigation reported in the body of the text, $D(x) = D_0$, $P(x) = P_0$ and $\Delta\rho(x) = \Delta\rho_0$ so that no variation of parameters along the plate is invoked. The reason is that an elastic plate with constant parameters satisfies the observations to better than about 1%, so that there is no resolution left in the data to determine anything beyond constant values.

2. The Inverse Tomographic Procedure

If all parameters are known, vertical displacement as a function of horizontal distance from the hinge point can be calculated and a theoretical basement profile plotted. However, the values of the flexural rigidity D, lateral stress P, bending moment M_0, initial end-point load V_0, dip angle θ, and the length of the plate L, have to be inferred from present-day geological or geophysical data. Denote by p the vector of all parameters describing the system. For a given choice of vector p, the predicted behavior, $w_M(x,p)$ will be determined at present-day. However direct measurements are available for $w(x_i)$ ($i=1, \ldots R$), where the vertical coordinate reference is the same as for the forward run model calculation. Clearly, with arbitrary choices of the vector p there is no reason why model predictions $w_M(x_i;p)$ and observations $w(x_i)$ at each location x_i should be similar. The aim is to use the difference between observed and predicted behaviors to provide a procedure which iteratively corrects the model parameters towards values leading to the least discord between predictions and observations. Such a procedure operates as follows. The parameters in the vector p have different dimensions. Let the jth component of p have a maximum value $\max_j$ and a minimum value $\min_j$. Then set

$$a_j = (p_j - \min_j) / (\max_j - \min_j) \tag{A13a}$$

so that $0 \leq a_j \leq 1$. Regard the vector a as fundamental with p being a dependent vector given through

$$p_j = \min_j + a_j(\max_j - \min_j) \tag{A13b}$$

In this way all components of a are dimensionless and are required to be in the range zero to unity.
Introduce the least squares control function

$$X^2(a) = R^{-1} \sum_{i=1}^{R} (w_M(x_i,a) - w(x_i))^2 \tag{A14}$$

An iteration scheme which <u>guarantees</u> to produce a closer correspondence of predicted and observed bed locations at each iteration, and which <u>guarantees</u> to keep each component of a in the bounds $0 \leq a_j \leq 1$ is:
At the nth iteration represent the updated value $a_j(n+1)$ to a_j by

$$a_j(n) = \sin^2\theta_j(n)$$

$$\theta_j(n+1) = \theta_j(n) \exp\left[-\tanh\left\{\alpha_j\delta_j(n)\frac{\partial X^2(a(n))}{\partial a_j(n)}\right\}\right] \tag{A15a}$$

$$a_j(n+1) = \sin^2\theta_j(n+1)$$

where

$$\alpha_j = \left|\frac{\partial X^2(a(0))}{\partial a_j(0)}\right|^{-1} \ln\{1 + (Ma_j(0))^{-1}\} \tag{A15b}$$

$$\delta_j(n) = q_j(n) / \left[\frac{1}{Q}\sum_{p=1}^{Q} q_p(n)\right] \tag{A15c}$$

with

$$q_j(n) = (|a_j(n) - a_j(n-1)| / a_j(n)) + \beta^2 \tag{A15d}$$

where M is the number of times the non-linear iteration is to be undertaken, $a_j(0)$ is the initial choice for a_j (with $0 \leq a_j(0) \leq 1$), Q is the number of parameters being varied and β is the scale-choice for calculating numerically the derivative $\partial X^2/\partial a_j$ (see below). The factor $\delta_j(n)$ chooses the most sensitive parameters for more rapid variation after the first iteration ($n \geq 2$) while for $n = 1$, $\delta_j(1)$ is set to unity.

This procedure guarantees to increase (decrease) a_j according as $\partial X^2/\partial a_j$ is negative (positive), guarantees to keep each component of a in the range $0 \leq a_j \leq 1$ and guarantees to make the largest changes in those parameters which are furthest from satisfying the least squares control criterion.

Two caveats are: (i) the numerical accuracy with which a minimum in least squares mis-match can be achieved is limited by the accuracy with which the partial derivatives, $\partial X^2/\partial a$, are calculated; (ii) the fact that the non-linear iteration scheme guarantees to find a minimum least squares mis-match is no guarantee that the procedure has found a global minimum.

The first caveat can be handled to an extent by decreasing the range of a particular component of the vector a used to calculate a derivative: thus with the approximation

$$\partial X^2/\partial a_i = (\beta a_i)^{-1} \{X^2(a_1, a_2 \ldots., a_i + \beta a_i, a_{i+1}, \ldots) - X^2(a_1, a_2 \ldots., a_i, a_{i+1}, \ldots)\}$$

and the default value $\beta = 0.1$, one can in turn replace β by 10^{-2}, 10^{-3}, ..., until some pre-set criterion is reached for accuracy of the derivative calculation. By and large this problem is not severe. As a by-product, if numerical resolution allows $a_i(n)$ to exceed unity, then it is an appropriate strategy to immediately set $a_i(n) = 0.9$, which forces $a_i(n)$ to be in the pre-set domain $\{0,1\}$, and which is also a valid value from which to calculate the partial derivative because, then, $a_i + 0.1 \, a_i = 0.99$, which is also in the required domain.

The second caveat is of greater concern because some procedure must be found to decide whether multiple minima exist, and to decide which minimum is the appropriate value to use in parameter determination, from which one constructs the combined paleo-evolution of the basement and associated sedimentary beds. A strategy which has proven useful in other similar problems is as follows.

First make a rough assessment of the minimum and maximum ranges that one anticipates will encompass each parameter. Then do a linear search over pairs of parameter values while holding the remaining parameters at their mid-point values. Two-dimensional plots of mean square mis-match contours in each two-parameter space then provide a rough idea of where minima lie, as well as suggesting how to modify the initial estimated ranges of parameter values in order to improve the ability to focus on the minima uncovered.

Then a multi-dimension non-linear iterative search, as outlined above, can be done to improve the degree of fit between predicted and observed behaviors. The point here is that all parameters are allowed to vary simultaneously (rather than just two) thereby improving the overall parameter values. A linear search can be redone after the non-linear iterations, so that one is more sure that a global minimum in the search ranges specified has been obtained. The pair of intertwined procedures (linear search, non-linear iteration) can be repeated as many times as desired. The improvement in least squares mismatch using this type of strategy is considerable, as well as helping to ensure that the search procedure does not become "stuck" in a local minimum.

10Mud Island and Mud Diapir Motion Hazards

In addition to the major hazards developed in the preceding chapters, there are also less frequently occurring (or less often recorded!) hazards associated with mud islands and diapirs. Historically people have observed that some of the islands in the Caspian Sea suddenly disappeared, or appeared again. Many islands appeared for a couple of weeks, or even days, and then were "washed out" by waves. Later people noticed that, most of the time, the appearance of new islands was associated with offshore eruptions. Therefore it was the opinion, for a long time, that the appearance of those islands was due to erupted mud, which constituted the islands. Probably in some of the cases this phenomena really did occur. But analysis of the observations shows that other processes take place as well. The appearance and disappearance of islands can be due to mud diapiric motion.

Here we give descriptions of these events so that one has some idea of the sorts of hazards they can represent. When a better statistical data base becomes available than is currently the case, it should be possible to provide probabilistic values of hazard assessments. However, at the present time of writing, about the best that one can do is to note the hazard types. But first a presentation is given of some summaries of historical observations of mud island occurrences and disappearances.

A. Historical Observations of Mud Island Occurrences

One of the ancient records of such phenomena was connected with the "Old Fort" in the 14th century. This fort, named Bail Castle, was built in the 13th century on the sea coast land to the South of Baku harbor as a protection for Baku city. This fort/trading post survived until the 14th century when a major earthquake occurred (or possibly it was an eruption of a mud volcano similar to the Hamamdag volcano eruption described in Chapter 3). In any event a huge fracture formed, which cut-off the castle from the mainland. The newly formed "island" then slowly disappeared beneath the sea.

The "island" resurfaced slightly in the mid-1950's (a 2 m drop in sea-level from 1930 to 1950 may have been a contributory cause or the mud island may have risen independently of the sea-level fall), when several of the fort's carved limestone blocks were recovered. These blocks showed the effects of several centuries of immersion and tidal blurring of parts of the carvings on some of the blocks (and can be seen today at the castle in Baku old-town). The island then disappeared again, maybe due to the rise of the Caspian Sea.

283

But consider now more modern, and more confident, data. One of the interesting islands is Chigil-Deniz (the old name was Kumani Island). Accordingly to the records catalogued by Dadashev (1995) an island was situated there. In 1860 the island was above sea-level by 1.2 m, and then sank into a submarine bank. The next eruption after this record happened the next year, May 7, 1861, and, according to Khalilbeili and Gasanov (1960) and Yakubov and Putkaradze (1951), the island was visited by Abikh, who measured the size of the "new" island to be 87 m x 66 m and height 3.6 m above sea level. The next recorded eruption occurred in May 1927, but between those two eruptions the island appeared a couple of times. Dadashev (1995) notes:

1862 - Above sea-level by 1.2 m.,
1869 - Above sea-level by 4.3 m.,
1910 - Above sea-level by 5.5 m.

After the eruption of 1927, the new island was visited by Abramovich (Yakubov and Putkaradze, 1951) and its size was 73 m x 55 m x 0.6 m. That island then washed out in a month. The next year (Yakubov and Putkaradze, 1951) the island appeared again with a diameter of 135 m and height 1.5 m. However in the next paragraph of their paper Yakubov and Putkaradze (1951) note that Kucenko measured an island twice and its radius was 144 m (1928) and 7 m (1929). In 1939 an island of 324 m x 290 m x 3 m appeared and washed out again.

An eruption of Dec. 4, 1950 was the strongest and was described by Yakubov and Putkaradze (1951), who visited the island the second day after the eruption.

Measurements in 1947 showed that the water depth was 7 m. On Dec. 4, 1950 at 18.40 an eruption of the volcano occurred. The eruption started with a bright flash. The flame height was over 100 m, and was observed from Baku. The color of the flame was orange. The flame in the lower part was wider and in 15 minutes the flame stopped instantaneously. Black smoke was in the sky for a long time after that. A new island appeared. The island consisted of semi-liquid mud with a strip of turbid water of 50 m width surrounding the island. The whole island was semi-liquid. So people used wooden boards to move about on the island.

The island had a length of 700 m and width of 500 m, and consisted of erupted mud. The long axis was directed NW-SE. On April 30, 1951 a Caspian Expedition of the Air Methods Lab of the USSR Academy of Sciences made an air survey of the island. The size of the island was reduced to 320 m x 230 m.

The next air photo survey was done on September 11, 1951. The island had a direction from NE to SW with a long axis of 200 m and a short axis of 110 m. The next air photo picture was made on June 3, 1952, when the remaining part of the island of 70 x 40 m^2 had a wedge-shape with a very sharp south-west edge. Zubenko visited the island in July 27, 1952 and found just a small piece of land of 15 x 5 m^2 with a height 1.7 m above sea level.

Returning again in August 3, 1952 Zubenko found that the island had completely washed out and become an underwater bank. The results of a photometric survey, as well as the description of the next volcano, were given in Khalilbeili and Gasanov (1960).

The eruption of December 25, 1959 was associated with a strong tremor and rumble, and a flame of 200 meter height. The eruption lasted 15-20 minutes and, as a result, an island of NW-SE direction was formed. The island had a length of 200 m and a width of 170 m, and was up to 2-2.5 m above sea level. The eruption center was in the central part of the island and periodically the ejection of gas and mud was observed to a height of 3-4 meters. According to Khalilbeili and Gasanov (1960), the island consisted of mud volcanic breccia slightly dried on the surface, with shallow and narrow fractures. But at the same time they noticed that the north-west and south-east coastal parts of the island were covered by large numbers of gryphons which emitted gas and liquid mud. This fact suggests that not all the island was covered by mud, otherwise the gryphons would have been covered too.

Another semi-permanent island in the Baku Archipelago is Kornilova-Pavlova. According to Yakubov et al (1972) the island was detected in 1903 for the first time, with a shape like a hill, the top of which was in a water depth of 3 m, and the base 40 m. In 1907 the depth of water above this bank decreased, and a boulder rose above the sea by up to 0.3 m. In 1910 the depth of water increased. On July 15, 1915 at this location an island of size 64 x 20 m^2 and height 1.5 m formed. Later (1921) this island disappeared and in its place a bank formed again. The depth of water that time was 6.3 m. In 1959 intensive gas and mud emission was recorded. A sonic survey showed that a funnel-shaped crater of a diameter 12-15 meters formed. This crater looked like a salsa does on the onshore volcanoes.

On June 25, 1970 Caspian sailors reported that a new island formed on the Kornilova-Pavlova Bank, 90 km to the south from Baku. The size of this island was length 105 m, width 65 m and height 1.5 m above sea level. The next day a geological expedition from "Azermorrazvedka" ("Azer-offshore-exploration") and the Geological Institute of the Academy of Sciences visited the island. More accurate measurements showed that the length of the island was 114 m and the maximal width 49 m with a direction NW-SE. The island consisted of 2 parts. The north-west part had a round form, while the south-east part had an asymmetric shape and looked like a jackboot with pointed toes. The parts were connected by an isthmus 12-14 m wide. The north-west part of the island consisted of terrigeneous detrital well-sorted rocks (In general boulders up to 1.7 x 0.8 x 1 m^3 and pebbles). The gravel with pebbles was raised over 1 meter above sea-level. On the crest of the island there were 4 gryphons (Yakubov et al., 1972).

The south-east part of the island was about 0.5-0.7 m above the sea and consisted of erupted mud, covered by pebbles and boulders. On the south-east edge there was a large boulder raised approximately 2 meters above sea-level. On this part of the island the erupted mud (breccia) was hard enough to stand on. The height of the new volcano was 6-6.5 meters. The eruption (?) was very quiet without the ejection of significant amounts of mud and gas. Just mud was steadily pushed out, which led to the raising of the sea-bottom. Only a small amount of mud flowed out to the surface. The island did not change its shape for a couple of months.

Makarov Bank volcano was already mentioned earlier. This volcano is to the north from the above mentioned volcanoes of the Baku Archipelago and, according to Sultanov and Agabekov (1959), during the 20th century eruptions on the Makarov Bank were recorded in 1906, 1912, 1917, 1921, 1925, 1933, 1941 and 1958. For the previous century the eruption of 1876 is known. The eruptions were associated with the ejection of gas and mud. The cones which formed during eruption sometimes rose a couple of meters higher than the sea surface and formed islands, which then disappeared as the sea currents washed out these masses of mud until the base of current flow was reached, leading to a submarine bank. This happened in 1876, 1921 and during the last eruption of 15 October 1958 which was described in Chapter 4. The next day after that eruption a group from the Academy of Sciences visited this place. The water was turbid over a significant area, which indicated an intensive wash-out process. The group did not see the island, but in their opinion the eruption could happen only from an island, because only in such a case could the sand particles be scattered in the air. As was shown in Chapter 8, it is extremely difficult to throw solid particles through the water column.

The next interesting volcanic island was described by Azizbekov and Yakubov (1938):

"The coordinates of Livanov Bank are 39°45' north and 52°6' east. It is near the ship track between Baku - Krasnovodsk. The Bank was discovered for the first time in 1895 after an earthquake. In 1896 the water depth here was 0.6 m, while the depth of surrounding water was 73.2 m. In 1898 the water depth here was 0.3m, and in 1903 was 5.5m. In 1908 the bank could not be found. The minimal water depth in this area was 49.4 m. In February 1928 and December 1929 the captains of ships reported bow-waves in the Livanov area. A commander of "Murav'yev" ship on September 21, 1930 saw a big fountain throwing water 5 meters above the waves. The surrounding water was clear. Two days later the ship "Kommuna" found an island of 2-3 meter height with a sand-bar spreading to the north-east to 200-300 meters.

A hydrographical expedition visited the island on the ship "Maksim Gorkii" on October 6, 1930. The island had a length not more than 40 m, width 20 m, height 1m. The edges of the island were very eroded by the water (washed away). The surface of the island was rough. The ground was sticky silt, but the surface was crusted. Near the island gas bubbles were visible in many places in the ocean. On October 13, 1931, according to the report of the commander of "Maksim Gorkii", the island did not exist any more. Also the bow-waves were not observed. From that time until 1937 there was no evidence of volcanic activity. On the maps of 1935 the Bank was indicated as being underwater with a water depth of 26 m from the east side and 73-76 meters from the west side. In 1937 reports about a new island came in March-April. On May 4, 1937 an expedition on the "Tbilisi" ship visited the island, which had a length 40 m, width 25 m and height 3.5 m. In the coastal zone of the island fragments of different rocks lay, while the inner part of the island consisted of cracked erupted mud."

Some island motions have been observed to the north of Apsheron peninsula. For example Zuber (1923) wrote:

"In winter of 1918-1919 I heard from one old man called Peters that about 30 years ago near Mardakyan village (north of Apsheron) an island appeared and then disappeared. He gave me a piece of sandstone from this island. Capitain Antonov testified that such a phenomena took place."

Zuber did not show exact coordinates of this island (probably he did not know then), but possibly this temporary island was Buzovninskaya Sopka, which is located between Buzovny and Mardakyan villages. Eruptions of that volcano have been observed a couple of times.

It was recorded in 1892 near Buzovny village (4-5 km from the shoreline) that an island formed, which later washed out and remained a bank with an overlying 2 meter water depth. At the same place the island appeared again in 1923, disappeared the same year and remained a bank with changeable water depth. The measurements in 1936 did not indicate any bank here, but on the sea surface people could see a lot of gas bubbles (Putkaradze et al., 1954). In 1950 a small island appeared without visible eruption (Ismailzadeh, 1954).

In 1953 the volcano erupted twice according to Putkaradze et al. (1954). The first eruption was on February 26, 1953. The drillers heard a muffled explosion sound at 8 p.m. There was no other evidence like flame, rock throwing, etc. The next day they saw a new small island. The eruption continued as a periodic throwing of rock and mud fragments. Periodically (every 40-60 minutes) the volcano ejected mud and threw small pieces of rock to heights of 15-20 meters. The same day at 4 p.m. the eruption stopped. On March 1 geologists came to the island and sampled the rock fragments.

The height of the island was 3-4 m and the size 60 m x 40 m. The surface was covered by mud breccia. The island washed out very intensively and disappeared in 20 days.

A second eruption occurred on September 10, 1953. As in the previous case it was not a spectacular manifestation of eruption. The second island was larger than the first, with a size of 100 m x 70 m and height 4-5 m above sea level. The mud of the second eruption consisted of more hard rock fragments than the first. The second island lasted 2 months. Approximate estimation gives the volumes of erupted mud to be 50,000 m^3 and 120,000 m^3, respectively, for the two eruptions.

B. Mud Diapirs and Semi-permanent Islands

Island appearances/disappearances are not a very common occurrence but are also not an unusual phenomena in the South Caspian. There is no doubt that such islands are related to mud diapirs or volcanoes. But the mechanisms by which they are manifested are different. Islands can be classified as following:

1. Evanescent Mud Islands

Often, after an offshore eruption, one is left with an island of mud, which can be up to several hundreds of meters in lateral extent and a few meters above sea-level. Over the course of a few days or weeks, such evanescent islands are washed away by the action of wind and waves. Such islands usually appear in places where the water depth is shallow. From Chapter 5 we know that the average thickness of mud flows is about 5-10 meters, which assumes that maximal thickness of the flow barely exceeds 20-25 meters. The results of the dynamical modelling of the flows, run in a number of cases, indicates the same scale of values. However, the results of drilling show that significant mud accumulations exist.

For instance, well No 20 on Bulla Island (Sultanov et al., 1967) crossed a layer of paleo-mud flow with thickness 300 meters, and well N22 one of 340 meters. Such paleo-mud covers (up to 130 m thick) are found around Bulla Island to a radius of 3 km. At increasing distance from the island those layers decrease in thickness and then disappear. However, it is doubtful that such layers could be formed during a single eruption. According to descriptions of eye-witnesses (Yakubov and Putkaradze, 1951), an island in the Kumani area in 1950 appeared as a result of erupted mud. As was mentioned in the previous section, the whole island was covered by mud. Other evidence was that a flame burned at the beginning of the eruption, stopped instantaneously rather than slowly, indicative of burning gas coming through water. Most likely, the eruption started from a submarine bank, where the depth of water was shallow (5-10 meters). Therefore, ignited gases had enough pressure to blast through the overlying water. When the pressure dropped lower than a threshold (equal to the hydrostatic pressure of the water column over the volcanic crest) the flame stopped instantaneously; but the eruption continued. Erupted mud then soon formed a cone which rose out of the water. So an island formed, which later was washed away by sea waves.

2. Mud Islands formed due to mud diapir rise

Diapirs are geological bodies filled by semi-liquid mud and are highly saturated with gas. In some conditions the gas goes from liquid solution to free-phase. In such cases the mud pushed by gas tries to find a vent to escape. But before that can happen the overlying sediments keep the mud-gas bodies buried. So such a mud body starts to rise like bread in an oven, and will continue to rise until the elastic limit of overlying sediments is exceeded and then, after fracturing, a volcano starts to erupt. Erupted mud releases the gas pressure and the crest of the diapir then slowly sinks.

Rising diapiric crests before eruptions have been observed many times on onshore and island volcanoes. For example, on Duvanny Island on September 4, 1964 the ground slowly lifted 10-15 meters before smoke, and then flame, appeared (Sulvanov and Dadashev, 1962). On Hamamdag volcano, the coast was lifted 10-15 meters (Agabekov and Bagir-zadeh, 1948). And after almost all onshore eruptions the craters sink (Chapter 3).

Kumani Island (in 1950) formed due to erupted mud. The description of the island formed in 1959 (Khalilbeili and Gasanov, 1960) allows one to suggest that an island arose before the eruption, then erupted and disappeared slowly; otherwise, gryphons, located at the coastline, would have been covered by mud too, as was the case in the center of the island. Two eruptions of

Buzovninskaya Sopka (Ismailzadeh, 1954; Putkaradze et al., 1954) happened the same way. People first saw in island then an eruption occurred. It is very hard to imagine that an island in the Livanova area appeared due to erupted mud. It is almost impossible to create a cone of 20-50 meters height in water (Azizbekov and Yakubov, 1938) and consisting of sloppy mud. So the appearance and disappearance of Livanova bank/island could be associated only with motion of the diapiric crest of the mud volcano.

3. Submarine/Aeolian Islands

Apart from the rise of mud diapirs to form islands, there is also a slower rise so that mud diapirs can sit with their crests below sea-level (banks). But below sea-level the crest of a diapir has an additional hydrostatic pressure. When rising above the water column this pressure drops to atmospheric pressure, therefore the chance of an eruption increases. However, we have too little information about such submarine eruptions, but an eruption of Makarov Bank in 1958 (Sultanov and Agabekov, 1959) was, most likely, from such a submarine bank.

Arguments presented by Sultanov and Agabekov (1959) that sand particles could be scattered due to air transport only from an island are not appropriate. Sand particles could be erupted through the water layer with burning gas and mud. As was shown in Chapter 6, the pressure can be enough to make a "hole" in the sea and to eject a significant amount of mud to a height of a couple of hundred meters.

4. Dry Push-Out Islands

As well as gas pressure driving the diapiric motion, diapiric mud can also be driven as a result of buoyancy or in other ways. This phenomena is called a "dry push-out" and was observed on Cheildag and Goturdag volcanoes (onshore). The report of Yakubov et al. (1972) states:

"The first eruption on Cheildag occurred in November 1870. The next one a hundred years later.

On July 4, 1970 at 7:30 p.m. a heavy rumble came from Cheildag mountain. Some time later a crash sound came and a flame appeared. We came to the volcano on June 10. The eruption was not usual. It continued for a long time. And only to the end of July did it finish. Instead of an eruption we observed that semi-dry mud was pushed out from a fracture of 60-70 meters length formed during the eruption. On the 9th day after the onset the velocity of flow of this mass was 1 m/day. Then it dropped to 65-40 cm/day.

The largest fractures had widths 0.5 m and depths 2-2.5 m. Burning gases came from those fractures and formed flames of 1.5-2m height. The walls of these fractures became red-hot. Gases burned also at depth below the erupted mud. About 3 months later the gases still burned.

As a result of the eruption a number of fractures formed. One of them extended to 1 km. The area of old eruption of 80-100 meters width and 400-

450 meters length was mostly fractured. It had a very rough topography. The blocks were lifted or fell for 2-3 meters, due to which the old crater came to the surface. It was filled by burned brick-colored mud, which implied underground gas burning. That happened in Eastern Cheildag. About 1 km away there is another volcano - Western Cheildag. From fracture of 15-20 cm width the mud was constantly pushed out. One can measure the velocity because every year after the winter rains the mud is stuck to the surface. So far during the last 4 years it moved with a speed of about 40 cm/year. In summer 1970 the speed increased to 1.4 m/year. This process is going until present "(which has been confirmed by measurements in 1992 during an expedition in which one of the authors (EB) took part). The same phenomena, but on a larger scale is observed on Koturdag volcano from the middle of last century. From the crater of 50 m x 15 m semi-dry mud is pushed out to flow down the slope. The rates of such a movement were 42m/year in 1926-27, 9.6m/year in 1937, 10m/year in 1949 and 1955, 18 m/year in 1959. This push-out process lasted more than 100 years. And in 1966 it finished with a strong eruption."

The question is: Does such a phenomenon occur in the offshore? There is not enough direct information to be sure, but possibly the appearance of Kornilova-Pavlova Island in 1970 without a "significant amount of mud and gas ejection" (Yakubov et al., 1972) was associated with such a process.

5. Almost permanent islands

There are islands in the Caspian Sea which are the tops of mud volcanoes. Some are known from ancient times (Bulla, Duvanny, Sangi-Mugan (Svinoi), Garasu (Los)); others appeared in the near-past; Kamen Ignatiya appeared in 1914 (Yakubov et al., 1972) and stayed until the Caspian Sea level rose in the late 1970-1980 period. Those island volcanoes are quite active. Eruptions are recorded on almost all of them. But there is no guarantee that one day they will not sink to the Caspian Sea bottom. The Caspian Basin is unpredictable.

6. Lateral Diapir Motion

In addition to the vertical rise of diapirs, and the flow of mud down the flanks of diapirs in both aeolian and submarine conditions, there is now considerable seismic evidence that there can be considerable lateral motion of a diapir - forming arcuate "scoop" faults - much as occurs in the Gulf of Mexico with shallow and/or unconsolidated sediments.

In short, while statistics of historical occurrences of such events are sparse, the fact that they occur at all is sufficient to give one pause to consider the hazard factor of their occurrence in exploration assessments.

C. Hazard Aspects of Island Motions

Three hazards are apparent during the existence lifetime of such evanescent islands: first, during the stage when the island initially appears after an eruption there is an immediate navigation hazard to supply boats; second, during the stage of being washed away there is a hazard to undersea pipelines and sub-sea completion equipment because breccia clasts, supported in the mud, are then freed to move; third there is a hazard after the island disappears because it can still have a shallow submarine crest below wave-base (a bank), again a shipping hazard. And there is always the hazard of re-occurrence of an eruption, of course, either during the lifetime of the island or after it has submerged once more.

Hazards associated with the rising mud diapirs are obvious. In addition to those hazards associated with evanescent islands, given the approximately 2 m rise in sea-level in the South Caspian in the last 20-25 years (Dadashev et al., 1995), there are two other significant hazards: one is the ability of many diapirs to have crests at, or just below sea-level; the other is the ability of many diapirs to convert contained gas to hydrate during sea-level rises. Thus a hazard to shipping and/or rig-siting, as well as a hazard to drilling, are created by submarine rises of mud diapirs, together with the bulk motion of submarine mud diapirs.

While the amount of seismic information delineating such motions is currently limited, it would seem that such effects are of considerable concern. After all, the idea of drilling a borehole and producing hydrocarbons therefrom, only to have the underlying mud diapir move laterally and so not only shear the borehole, but also cause major hydrocarbon leakage from the sheared domain, is a considerable hazard both to production loss and to environmental clean-up costs.

Of even more concern are the "permanent" islands. They are considered as permanent because they have existed 100 years or more and there are many facilities on those islands (Meteorological stations, power supply, rigs, etc.). If such islands, like Bulla or Sangi-Mugan, one day start to sink then there would be major economic losses.

11 Hazards of Variability in Reservoir Characteristics for South Caspian Oil Fields

A. Introduction

The oil fields of Azerbaijan, both onshore and offshore, have been in production for over 130 years. During that period records of varying quality have been kept concerning the attributes of individual boreholes, oil and gas production, and reservoirs. Many of the very early records are either extremely incomplete or lost, while those records prior to the second World War are dominantly cumulative summaries of yearly oil/gas production. During and since the second World War, however, extremely detailed records have been kept of hydrocarbon production and of reservoir formation properties. Indeed, many of the original field record books are still available and, what is more, so are many of the individuals who either kept the field data or added to the field data. Some of those individuals are still working in the area. Thus the ability exists not only to obtain the original field data but also to question individuals about specific entries. In addition, there are often discrepancies between the actual field data records, and the political formal reports sent to controlling authorities of the time.

We have had the good fortune to have obtained access to the original field reports (some handwritten and, later, some typed) in which a wealth of data exists concerning production statistics, reservoir properties, structural and stratigraphic trap types, etc. These records have been translated, the data checked and then loaded onto a modern computer. In this way it is a relatively simple exercise to persuade spread-sheet computer codes to generate statistical distributions for all information loaded. We have done so.

Each attribute that one chooses to investigate must first be referred back to the comments in the original field data notebooks to ensure its veracity. We have done this as well.

Some attributes are of a statistically more robust character than others, of course. For instance, the number of producing horizons per field, for both onshore and offshore fields, is clearly an extremely robust statistic and has been used (Bagirov and Lerche, 1998) to predict the likely occurrence of future oil fields. Also for instance, the spatial locations of the producing oil fields are known and so the spatial distribution of producing horizons per field can be used to suggest likely regions for high finding probabilities for hydrocarbons

293

and to tie-in the observed distributions to the tectonic evolution and sediment supply for the South Caspian Basin since Miocene time after Tethys closure (Bagirov, Bagirov and Lerche, 1998).

This chapter considers a different group of statistical properties, also of a robust nature, but more directly related to reservoir characterization for the Basin.

From the available data base we have extracted both the mean producing thickness of each horizon for each field, the permeability, the porosity, etc. These data have been analyzed for frequency of occurrence, variations between onshore and offshore fields, and for a variety of other statistical contributions to delineate the range of likely values to be encountered.

This information is useful in that when one must perform an assessment of likely reserves in an undrilled prospect one has to have available some way of characterizing the reservoir volume, thickness, permeability, porosity, etc., and one must also have some idea of the relative uncertainty of each of the parameters entering such a reserve calculation. Thus a hazard statement is then available for hydrocarbon reservoir probabilities.

One such devise for aiding in this quest is the statistical patterns of behavior based on historical precedent.

B. Specific Distributions of Reservoir Characteristics

The data base consists of information for 37 onshore and 18 offshore oil fields that either have produced in the past or are still producing. A listing of the oil fields used in this paper and the number of producing horizons per field is given in Appendix A.

This section of the chapter considers two classes of reservoir distribution characteristics: geometrical and physical.

The geometrical distributions refer to depths, areas, thickness and volumes; the physical distributions refer primarily to permeability and porosity. Both classes are considered for onshore and offshore groups of fields taken together and also separately. Consider each class in turn.

1. Geometrical Distributions

a. Combined Onshore and Offshore Fields

(i) Producing Thickness

Inspection of Figure 11.1a shows that the total distribution (for all onshore and offshore fields used in the study) is very closely described by a log normal distribution with a peak value of around 15 m thickness. There is a long "tail",

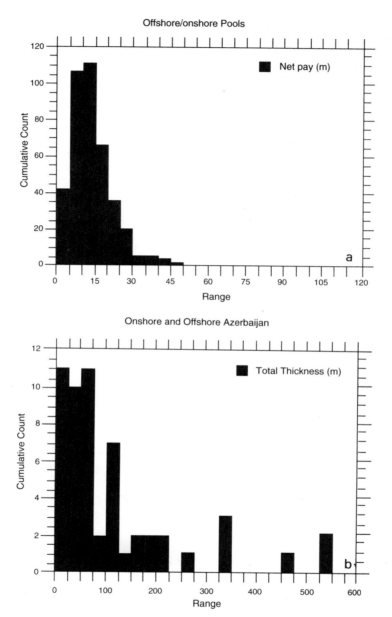

Figure 11.1. Combined onshore and offshore fields: (a) Net pay thickness distribution; (b) Total reservoir thickness distribution.

reaching to around 50 m thickness, but the distribution is dominated by producing thickness between about 5-30 m.

296

By way of comparison, plotted on figure 11.1b is the distribution of <u>total</u> thicknesses for the producing formations, showing a dominance of values at less than 100 m but with a long "tail" extending to around 550 m. Thus the typical production thickness of around 15 m is around 15-30% of the total thickness for formations less than about 100 m thick, dropping to around 8% for very thick formations.

(ii) Areal Distribution

Here the area mapped out by the total number of producing wells per field is provided. Inspection of figure 11.2, for both the onshore and offshore fields combined, shows that the distribution is effectively exponential with a mean value of around $8 \times 10^6 m^2$.

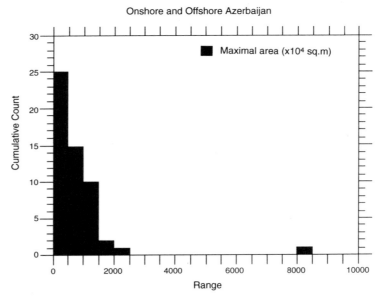

Figure 11.2. Combined onshore and offshore fields distribution of reservoir areas.

(iii) Volumetric Distribution

The producing volume of each field is portrayed in figure 11.3, determined as the product of area and average producing thickness. What is clear from figure 11.3 is that, once again, the distribution of volumes is dominated by the large number of small fields - as expected based on information from other areas of the world (Doré et al., 1996). There are only 6 fields with a volume of reserves in excess of $10^9 m^3$, and 44 with lesser volumes. The lower grouping has an

average volume of around $2 \times 10^8 m^3$ and, once again, is approximately
described by an exponential shape.

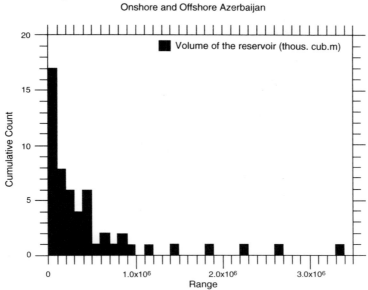

Figure 11.3. Combined onshore and offshore fields distribution of reservoir
volumes.

(iv) Reservoir Depth Distribution

In addition to the actual geometrical characteristics of the reservoirs, also of
concern is the typical sort of depths one would have to drill to in order to
encounter hydrocarbons. Shown on figure 11.4 is the distribution of average
depths to each producing reservoir. Note that the peak of the distribution is
around 1000 m, that the distribution is effectively log normally distributed, with
a tail stretching to around 6500 m depth, although the majority of the reservoirs
are shallower than 2500 m. Thus the sub-sediment depth to drill to is, in
general, quite shallow.

b. Onshore or Offshore Geometric Distributions

Similar groups of figures (as for the combined statistics) have been derived for
the onshore and offshore fields taken separately. Thus, figure 11.5a shows the
average total thickness for onshore fields alone, while figure 11.5b shows the
corresponding thickness distribution for the offshore fields alone. Inspection of
both components of figure 11.5 shows that both are dominated by values under
100 m thick, with only a small percentage of reservoirs greater than 100 m
thick. To within the limited statistics there is no discernible difference, of a
statistically robust character, between the two distributions for onshore and
offshore.

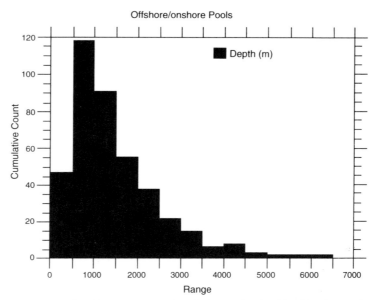

Figure 11.4. Combined onshore and offshore fields distribution of reservoir depths below the sediment surface.

Also, drawn on figure 11.6a (onshore fields) and figure 11.6b (offshore fields) are the corresponding areal distributions, showing that the onshore fields (figure 11.6a) have a fairly scattered distribution, ranging up to around 2.3 x $10^7 m^2$ with an average around 5 x $10^6 m^2$. The offshore fields on the other hand range up to 8 x $10^7 m^2$ but are dominated by field areas typically of order 1.0-1.5 x $10^7 m^2$. Thus, areas of offshore fields are on average 2 to 3 times as large as onshore fields.

Figure 11.7a (onshore fields) and figure 11.7b (offshore fields) show the corresponding volumetric values, again illustrating that the typical volumes of onshore reservoirs are about 5 x $10^8 m^3$ relative to the typical 3-4 x $10^8 m^3$ for offshore fields. There seems to be very little difference between the two groups based on the currently available statistics as exhibited in figures 11.7a and 11.7b.

Similar patterns of behavior are noted for depths to horizons as exhibited in figure 11.8a (onshore) and figure 11.8b (offshore) - both show typical sedimentary depths of around 1000-3000 m with, perhaps, a longer and significant "tail" at higher depths for the offshore case - although it would be difficult to make an unequivocal statistically sound argument at this stage.

2. Physical Parameter Distributions

In addition to the geometric considerations, downhole information, together with cumulative production information, indicates both the likely distribution of parameters influencing reserve and producible estimates, as well as the ranges of such parameters.

a. Porosity

Figure 11.9a presents the porosity distribution for the onshore producing horizons, indicating a typical porosity of around 17-25%, while the corresponding porosity distribution for offshore fields, depicted in figure 11.9b, shows a broader range varying from around 10-28% but with a peak value of around 23% - very close to the peak for the onshore statistics.

There is very little correlation of porosity with increasing sedimentary burial depth for either the onshore (figure 11.10a) or offshore (figure 11.10b) statistics. Certainly one could argue for a slight decrease in porosity with increasing burial depth for the offshore statistics but deeper than about 4000 m one is, perhaps, limited by the paucity of the statistics.

b. Permeability

The variations of permeability for the onshore and offshore fields are displayed in figures 11.11a (onshore) and 11.11b (offshore), respectively. There is a slightly larger mean permeability (~300 mD) for the offshore distribution than for the onshore situation (~200 mD), but both distributions are exceedingly broad, making the mean of less than statistically sharp worth.

Permeability versus porosity plots for the onshore data (figure 11.12a) and offshore data (figure 11.12b) show very little if any correlation so that there is then no discernible variation of permeability with burial depth either, as exemplified in figure 11.13a (onshore data) or figure 11.13b (offshore data).

c. Oil Viscosity

For both the onshore oils (figure 11.14a) and offshore oils (figure 11.14b), plots of oil viscosity distributions show that both are approximately exponentially distributed with a typical oil viscosity for onshore of around 7 mP/s and around 3 mP/s for offshore. There is no other discernible variation in the data base available at the present time of writing.

d. Reserves, Recovery Factors, and Water Content

For the combined onshore and offshore fields, figure 11.15 presents the total initial estimated reserves distribution, with figure 11.16 giving the total cumulative production and figure 11.17 the recovery factor, i.e. the cumulative produced amount of oil divided by the amount estimated to be present. One notes that the average recovery factor is around 20%, suggesting 80% of recoverable hydrocarbons have not yet been produced. There is a broad range of recovery factors but none exceeds 60%, and 80% of the recovery factors are less than 40%.

300

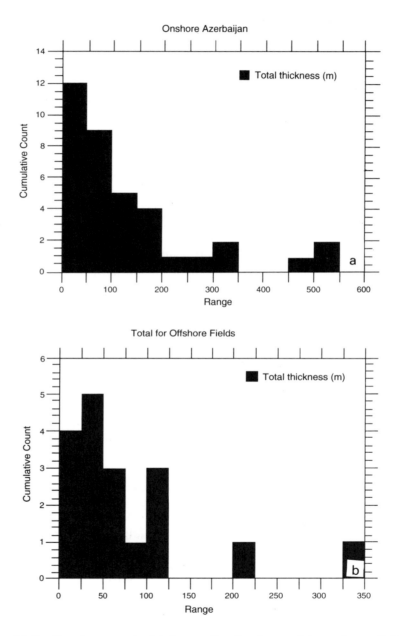

Figure 11.5. Total reservoir thickness distributions: (a) onshore fields; (b) offshore fields.

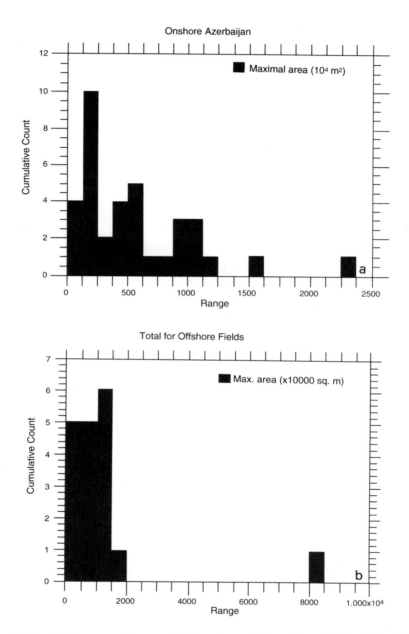

Figure 11.6. Total reservoir area distributions: (a) onshore fields; (b) offshore fields.

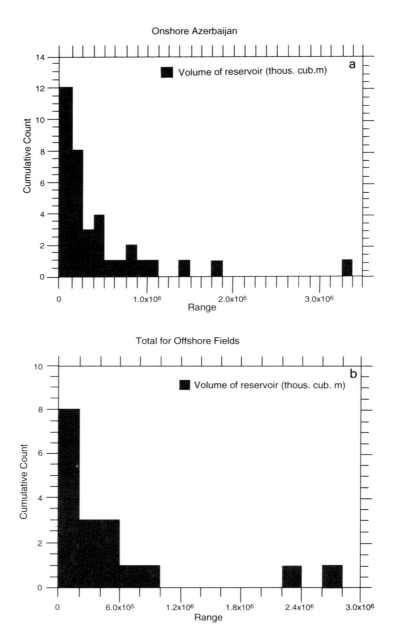

Figure 11.7. Total reservoir volume distributions: (a) onshore fields; (b) offshore fields.

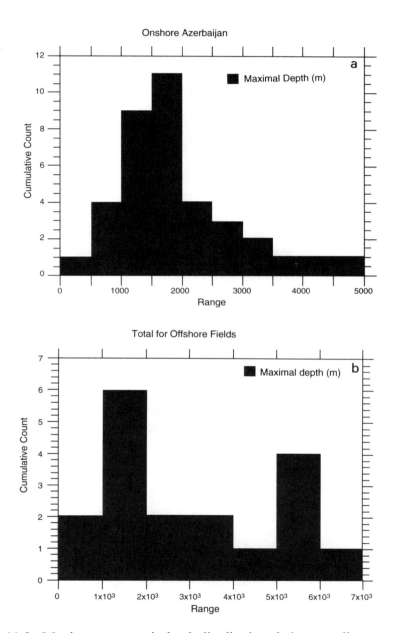

Figure 11.8. Maximum reservoir depth distributions below a sediment surface: (a) onshore fields; (b) offshore fields.

304

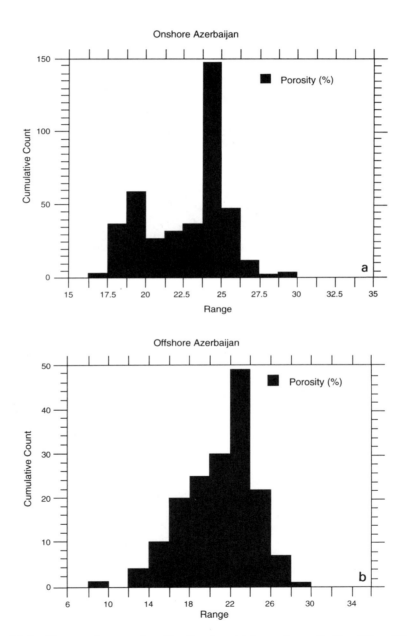

Figure 11.9. Reservoir porosity distribution: (a) onshore fields; (b) offshore fields.

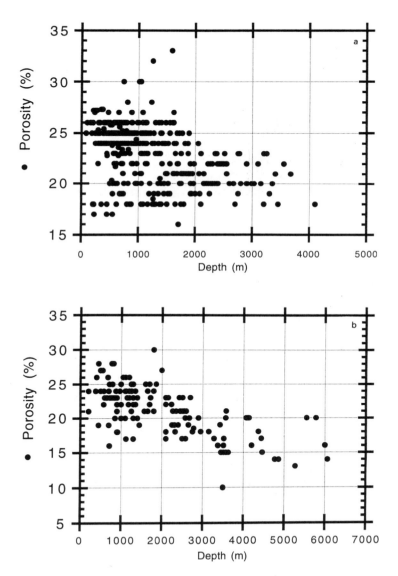

Figure 11.10. Reservoir porosity variations with depth below sediment surface:
(a) onshore fields; (b) offshore fields.

306

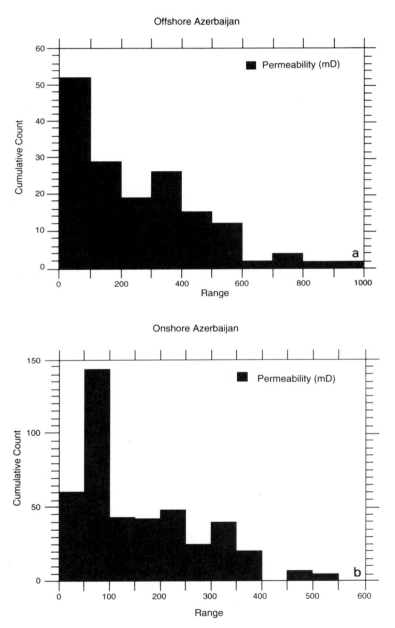

Figure 11.11. Reservoir permeability distributions: (a) onshore fields; (b) offshore fields.

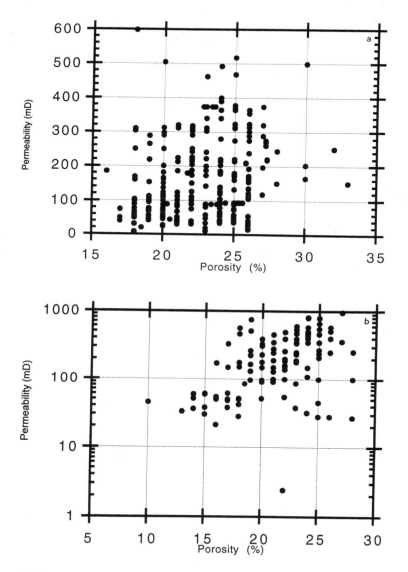

Figure 11.12. Cross-plot of reservoir permeability values versus porosity values: (a) onshore fields; (b) offshore fields.

The same pattern of behavior is present when one considers the onshore and offshore fields separately. Thus, figure 11.18a shows the initial assessed reserves for the onshore fields used in the study, while figure 11.18b

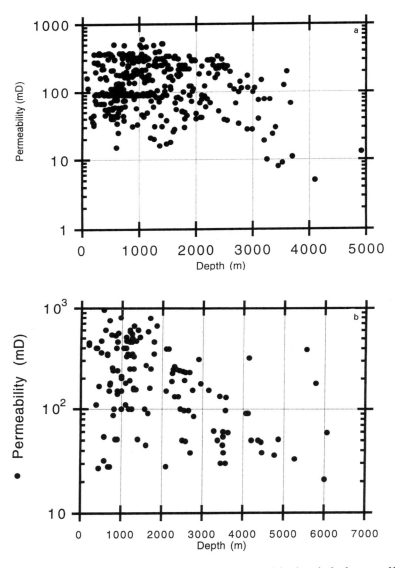

Figure 11.13. Reservoir permeability variations with depth below sediment surface: (a) onshore fields; (b) offshore fields.

shows the total cumulative production, with figure 11.18c giving the oil recovery factor, again indicating that significant reserves can still be produced because fractional recovery factors are around 0.3 ± 0.25. For the offshore fields the same patterns show through as exhibited in figure 11.19a for assessed reserves, figure 11.19b for cumulative production, and figure 11.19c for the recovery factor. Once again, the recovery factors are around 20-30% so that a significant amount of oil remains to be produced from the known offshore fields.

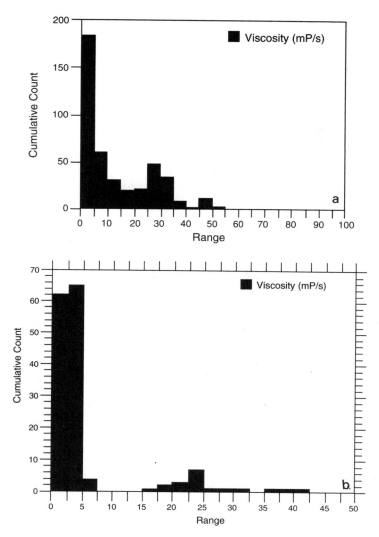

Figure 11.14. Distributions of oil viscosity: (a) onshore fields; (b) offshore fields.

However, it is not just the total fraction of hydrocarbons remaining which is of relevance. It is also important to know the water content per barrel of oil produced. This factor can play a significant role in evaluating the worth of continuing to produce. In addition, the fractional shale content in a reservoir sand is also of considerable importance because reservoir continuity (and so the ability to more easily produce residual reserves) can be impacted significantly by a high shale content.

Accordingly, from the available data base we have also derived the percentage water content in production and the percentage shale content in the reservoirs for both the onshore and offshore fields.

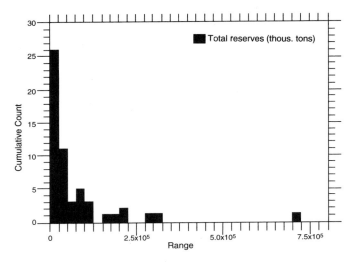

Figure 11.15. Total reserves distribution for combined onshore and offshore fields.

Figure 11.20a (onshore fields) and figure 11.20b (offshore fields) show the percent water content. To be noted here is a significant difference. For the onshore fields (as shown in figure 11.20a), the bulk have greater than 70% water in production, with some as high as 95-98% (i.e. 95 barrels of water for each 5 barrels of oil). For the offshore fields, the percent water content is more uniformly distributed, with an average
of around 50% water in production but with about half the fields smaller than 50% (figure 11.20b).

For the fractional shale content, the onshore fields have an approximately Gaussian distribution centered around 30% shale and with a half-width of ± 10% around this value (figure 11.21a), while the offshore fields have a broader distribution of shale content centered around 25% shale and ranging to around 10-50% shale (figure 11.21b).

e. Reserves versus horizons and depths

As part of the statistical data base we also have available the number of producing horizons per oil field and the mean depth to each horizon.

Accordingly it is possible to see whether there are any systematic patterns of behavior of reserves with horizon number or with depth to horizons.

Shown in figure 11.22a are the reserves for each field as a function of the number of producing horizons. There is some rough correlation in that the larger the number of producing horizons so, too, the larger the total reserves. But the scatter in this rough trend is considerable. For instance, the field with 20 producing horizons has 3×10^8 tons of reserves whereas the field with 43 producing horizons has only 2×10^8 tons, exemplifying the level of spread.

Shown in figure 11.22b is a plot of total reserves with maximal depth of each producing horizon indicating that there is no particular correlation - one is as likely to find high (or low) total reserves whether the producing horizon is shallow or deep.

f. Reserves versus reservoir thicknesses and areas

A cross-plot of total reserves versus reservoir thickness does indicate a strong correlation, as shown in figure 11.23a for the combined onshore and offshore statistical data, with (very roughly) a factor 30 increase in total reserves for every factor 10 increase in total thickness.

A corresponding cross-plot of total reserves versus maximal area of each field is given in figure 11.23b and also shows a clear trend with almost a 100-fold increase in total reserves for every factor 10 increase in maximal area.

g. Reserves and Production versus depth

For each field a cross-plot has been drawn (figure 11.24a) of the total initial reserves versus depth for both the onshore and offshore fields. To be noted is that the bulk of the reserves in both onshore and offshore cases are from fields shallower than about 3 km with very little of the reserves in excess of 3.5 km.

However, just because commercial oil is found at a given depth interval does not mean that it is also produced - for example, there can be shallower horizons in the same field which are easier to produce, or the water percentage may be considerably higher in one horizon than in others.

Accordingly, figure 11.24b exhibits the cumulative production for both the onshore and offshore fields, illustrating that the bulk of the production is from horizons shallower than 2.5 km, and with virtually no significant production greater than 3 km for the onshore fields, and for only a very limited number of offshore fields deeper than 3 km.

In short: the reserves in both the onshore and offshore cases are heavily weighted to less than 3 km, as is the production record.

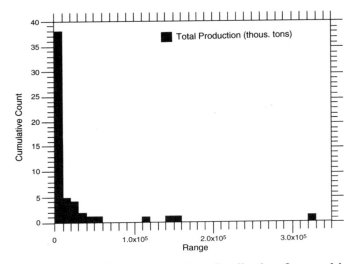

Figure 11.16. Total cumulative production distribution for combined onshore and offshore fields.

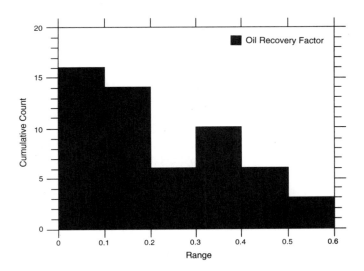

Figure 11.17. Oil recovery factor distribution for combined onshore and offshore fields.

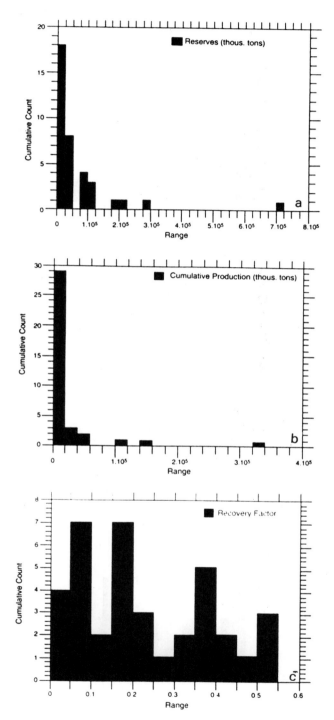

Figure 11.18. For onshore fields alone, distributions of: (a) total reserves; (b) cumulative production; (c) oil recovery factor.

314

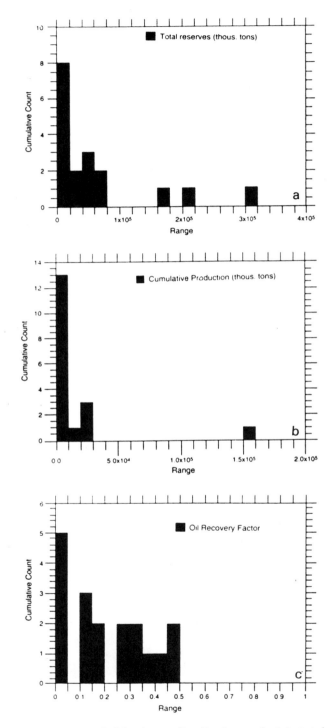

Figure 11.19. For offshore fields alone, distributions of: (a) total reserves; (b) cumulative production; (c) oil recovery factor.

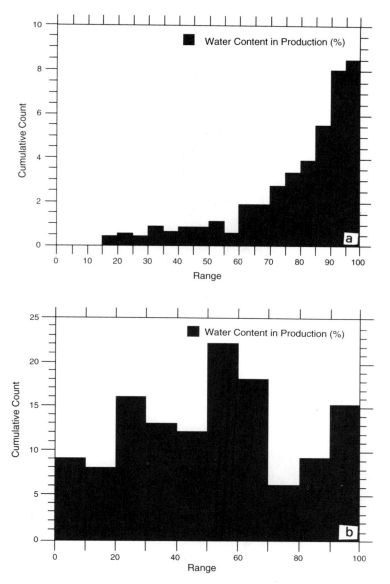

Figure 11.20. Distributions of water content (%) in production: (a) onshore fields; (b) offshore fields.

316

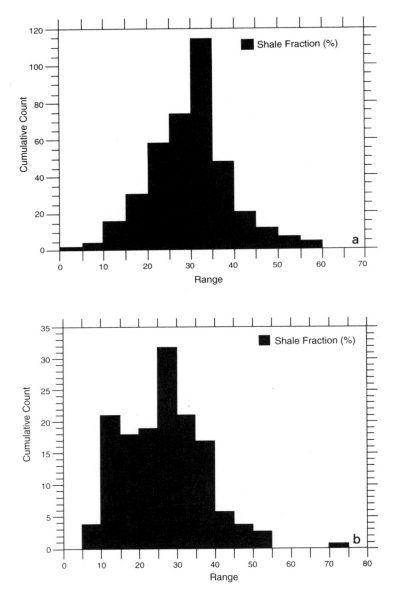

Figure 11.21. Distributions of shale content (%) in the reservoirs: (a) onshore fields; (b) offshore fields.

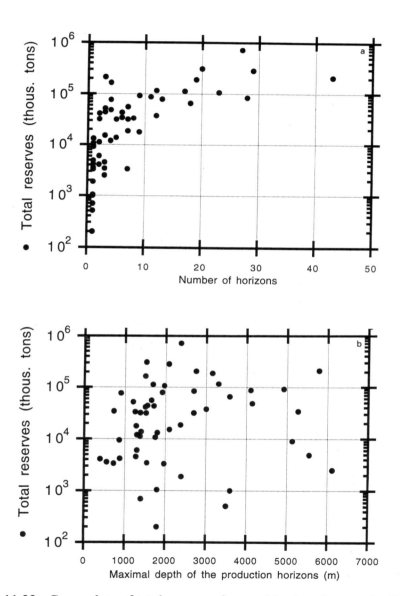

Figure 11.22. Cross-plots of total reserves for combined onshore and offshore fields versus: (a) number of producing horizons; (b) maximal depth of producing horizons.

318

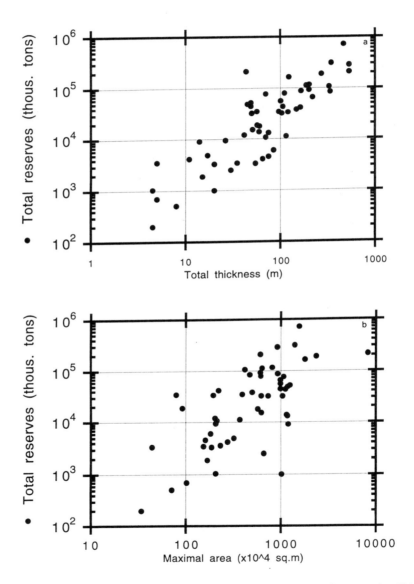

Figure 11.23. Cross-plots of total reserves for combined onshore and offshore fields versus: (a) total reservoir thickness; (b) maximal reservoir area.

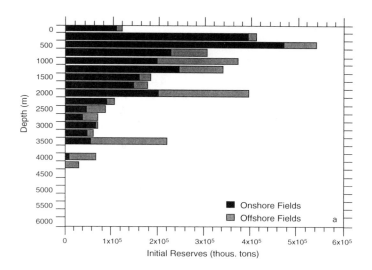

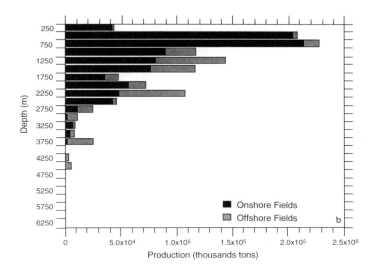

Figure 11.24. Cross-plots with depth below sediment surface for onshore and offshore fields of: (a) total reserves; (b) cumulative production.

C. Spatial Distribution of Economic Oil Field Horizons

While most of the Azerbaijan fields produce from the Productive Series (Plio-Pleistocene) formation, production is also found from formations as old as Cretaceous.

The number of productive horizons per field is also highly variable, ranging from 1 to 43, and each producing horizon also has variable depth ranges, sometimes over more than a 2-3 km range.

The oil types found are variable in sulfur content, in aromatic/aliphatic ratio, in density, in viscosity, etc., but there is, overall, a strong indication for a prevalent source rock (the Maikop (Miocene)), but not for an exclusive source rock (Guliev et al., 1998).

More details concerning individual fields and oils for each field and horizon are available (B. Bagirov, 1998), while consideration of oil field statistics is also available (Bagirov and Lerche, 1997). Summaries of the evolution of the South Caspian Basin are also collected in book form (Ali-Zadeh et al., 1996, 1997). The evolution of the Basin as a whole is complex, and mud diapir motion is still ongoing (Bagirov and Lerche, 1997), so that reservoir sands deposited since Tethys Ocean closure have been massively disturbed. Nevertheless, there is a sufficient amount of data on oil fields that it is possible to provide some statistical representation of likely locations for oil exploration.

The purpose of this section is to show how the number of horizons/field, together with field locations, can be used to enhance this oil-finding potential and so lower the hazard risk of not finding oil.

For this section our concern is only with the oil fields that either have produced in the past or are producing today. In Appendix A we present a listing of those fields together with the number of producing horizons per field. The number of producing horizons per field and the locations of each field are relatively firm numbers so that conclusions can be drawn based solely on these values.

Shown in figure 11.25 is the distribution of the onshore and offshore fields which were used in the analysis. Each field is represented by a blackened area, proportional to the interpreted producing area of each field, and each field has a number beside it which is its catalog number in Appendix A. Fields 8, 9, 10 and 14 are partially offshore and partially onshore, while 15 of the 46 fields used are totally offshore and 27 are totally onshore.

For each field, the number of producing horizons was entered into a spread-sheet and the contour map of figure 11.26 drawn, thus producing contours of economic horizons/field across the same area as for figure 11.25, and drawn at iso-contour intervals of 1 horizon/field.

Figure 11.27 superposes figures 11.25 and 11.26 to illustrate that the contours in figure 11.26 are not beholden to just one or two oil fields, but are drawn squarely through the middle of the distribution of oil fields. Inspection of figure 11.26 (or figure 11.27) indicates areas of likely high prospectivity from an oil-finding viewpoint.

Note, for instance, the broad valley of the low number of oil-bearing horizons/field running roughly NW-SE. Note also the two "bulges" of high numbers of horizons/field to the north-east and south-west.

The complex of high numbers of horizons/field along the Apsheron peninsula, both onshore and offshore, and continuing to the south-east of Baku Bay, provides a major trend in oil-field exploration potential. The "gap" between Ali-Bayramly and Neftchala (on the south-west trend of high number of horizons/field) is also a most likely zone for encountering structures with oil potential both onshore and, by extension, offshore to the south-east of Neftchala.

There are excellent geological reasons for this set of inferences from the data. First, the two main sources for sedimentary deposition after Miocene closure of the Tethys Ocean are from the north of the Basin and from the Kura depression to the west. Second, eastwardly directed thrusting from Talysh-Vandam and southerly-directed compression from the North Caspian, connected to the rise of the Greater Caucasus mountains, both serve to produce late-developed structures (Nadirov et al., 1996; Bagirov et al., 1997). Thus one anticipates that there should be at least a rough tie-in of the oil-bearing horizons/field to a combination of the tectonic events and to the sedimentation - as observed.

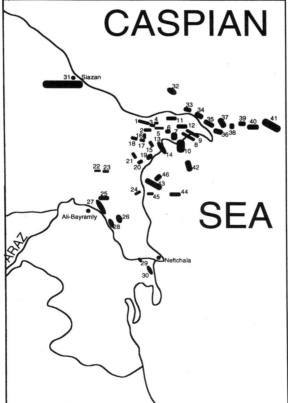

Figure 11.25. Map of locations of oil fields used in this study. Each field is located by a number referring to its position in Appendix A.

Of course, whether individual structures are oil-bearing or not depends also on local factors, such as faulting (and any associated leakage), on sand-continuity, and on the hydrocarbon charging ability of each sand. But the broad trend effect is clearly present.

The ability to extract inferences concerning potential hydrocarbon-prone areas in a basin is one which is often a critical ingredient in suggesting the types of fairways and structures for which one should be searching, thereby reducing the hazard of drilling dry structures.

In the case of the South Caspian Basin, the only "ground-truth" data on which to assess such potential hydrocarbon areas is the historical record. Because the number of economical oil-bearing horizons per field is a readily available value, that factor has been used here in an attempt to suggest likely areas for future exploration. Structures in such areas would have the highest chance of having the largest number of oil-bearing horizons, and so of minimizing the probability of dry hole drilling.

While not a universal panacea, nor a guarantee that one will find hydrocarbon-bearing structures, the number of horizons/field argument does provide a very likely general regional framework for more detailed investigations of local-scale structures that could be hydrocarbon-bearing. And this point is, perhaps, the most important one to make here.

D. Discussion and Conclusion

The purpose for compiling these statistics for the onshore and offshore South Caspian Basin is to facilitate the likely sort of reservoir sizes, volumes, physical attributes, and depth ranges where hydrocarbon occurrences can be found for exploration prospects.

The involvement of Western oil companies in the Caspian Basin in general, and the South Caspian Basin in particular, has surged over the last 8 years or so. The ability to evaluate likely hydrocarbon-bearing prospects in the Basin, and to assign some estimate of recoverable hydrocarbons to each such prospect, is a major concern to the oil industry. Both revitalization of old oil fields and exploration for new fields are ongoing. In efforts to aid these endeavors the historical data base of information provides at least some statistical guide to the ranges of behaviors to be expected.

Perhaps the dominant themes are that hydrocarbons are prevalent at less than around 3 km sub-sediment depth, that reservoir porosity is around 20%, that reservoir shale content is around 30%, and that reservoir permeability is around 200-300 mD.

There is some correlation of reservoir reserves with formation thicknesses and areas, but less correlation with other variables. Physical reservoir properties show little correlation with burial depth, nor are permeability and porosity that well correlated.

It is hoped that the information reported here will be of benefit on two fronts: first, to help quantify the likely reservoir properties as an aid to commercial oil estimates; second, to help those involved in studying the

Number of horizons

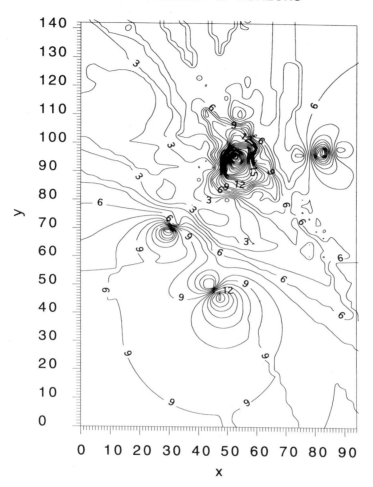

Figure 11.26. Iso-contours of number of oil-producing horizons/field across the study region. Contour spacing is unity.

dynamical, thermal and hydrocarbon generation, migration and accumulation properties of the Basin, by providing a compendium of statistical information against which models of basinal evolution can be compared and contrasted.

In these ways the historical data base compilations provide likely ranges of reservoir characteristics and so help to constrain hazards due to uncertain values in exploration ventures in the South Caspian Basin.

324

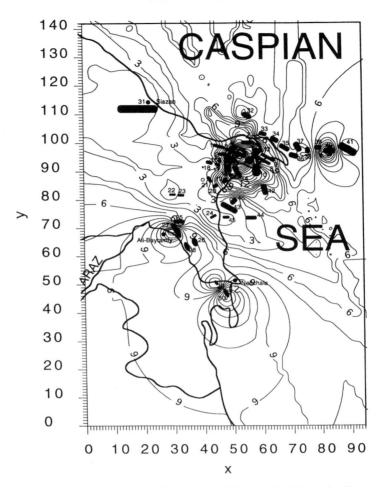

Figure 11.27. Superposition of figures 11.25 and 11.26 to indicate that the contours of figure 11.26 are not dominated by a single oil field, and also to relate likely regions for hydrocarbon exploration.

Appendix A - List of Onshore and Offshore Oil Fields used together with producing horizons/field

1 - Fatmai - 2
2 - Binagadi (north and south) - 14
3 - Kirmaki - 7
4 - Kurdakhany - 1
5 - Balakhany-Sabunchi-Ramany - 23
6 - Surakhany - 43
7 - Karachukhur - 28
8 - Zykh - 7

9 - Hovsany (Govsany) - 4
10- Gum Adasy (Ostrov Peschannyi) - 18
11- Buzovna-Mashtagi - 13
12- Gala (Kala), including Staryi Gala - 23
13- Bibi-Eibat - 29
14- Bukhta Ilicha - 12
15- Shabandag-Shubany-Yasamal Valey-Atashkah - 6
16- Sulu-tepe - 5
17- Shabandag - 1
18- Gara-Eibat - 1
19- Lokbatan-Puta-Kushkhana - 12
20- Kergez-Gyzyl-tepe - 4
21- Shongar - 1
22- Adjiveli - 1
23- Umbaki - 2
24- Dashgil - 1
25- Mishovdag - 3
26- Kursangi - 9
27- Kurovdag - 19
28- Garabagly - 11
29- Khylly - 3
30- Neftchala - 17
31- Siazan monocline, including: Chandagar-Zorat, Siazan, Saadan,
    Amirkhanly, Zagly, Tengialty - 11
32- Banka Apsheronskaya - 1
33- Banka Darvina - 3
34- Pirallahy (Artem) - 4
35- Gurgan-more - 3
36- Janub (Yuzhnaya) - 1
37- Chilov (Zhiloy) - 5
38- Azi Aslanov - 1
39- Palchyg Pilnilasi (Gryazevaya Sopka) - 7
40- Neft Dashlary (Neftyanyie Kamni) - 20
41- Gunashli - 4
42- Bahar - 6
43- Sangachal-Duvanny-Ostrov Bulla - 3
44- Bulla-more - 3
45- Alyat-more - 1
46- 8 Mart - 1

12 Natural Hazards in Hydrocarbon Exploration and Production Assessments

The overall purpose of this book has been to describe the main types of geological hazards present in the region of the South Caspian Basin, and to try to quantify, where possible, the probabilities of occurrence and the spatial scales of such catastrophic events. Using methods given in the body of the book one can calculate or estimate the chances of occurrence (or non-occurrence). Such events can lead to catastrophic financial losses for a company during the years of hydrocarbon exploration, development and production in an area of interest. As has been shown, an eruption of a mud volcano or associated flames, mud flows and hydrate explosions can destroy platforms, pipelines and other equipment. All of those factors should be kept in mind when decisions on exploration and exploitation are made.

Another prevalent concern that is becoming manifest in decision tree analyses for hydrocarbon exploration projects is with environmental clean-up if something goes wrong. For instance, if a project is saddled with the small chance of a catastrophic blow-out while drilling, and the associated high clean-up costs, then it is all too easy to take what could have been a very worthwhile exploration project and downgrade it to less than worthwhile. Equally, even if a blow-out scenario is not included while drilling, a similar catastrophic loss condition is often envisioned by management: economic oil (or gas) is found and, once into production, there is the chance of an oil spill with, again, enormous environmental clean-up costs. Accordingly, management often uses the catastrophic loss excuse to get out of (or never get into) good exploration projects.

Thus management bias can lead to a re-evaluation which takes, on average, a very good prospect and makes it seem very poor.

But the point about all of these management biases is that if only the expected value (average value) is used to quantify the risk, then one is completely overlooking the fact that there is also an uncertainty on the expected value, which ameliorates the "catastrophic loss" effect. The purpose of this chapter is to show how this amelioration operates in practice. Two case

327

histories will be outlined: "catastrophic failure" during (a) the exploration assessment; and (b) the development assessment.

A. Catastrophic Loss in Exploration Assessments

A blow-out (or other catastrophic failure) while exploring, due to a natural cause, will lead to loss of equipment and massive environmental clean-up costs. The classical way to assess catastrophic failure chance operates is as follows. First an evaluation of an opportunity is made, yielding an estimate of costs, C, potential gains, G, and chance, p_s, of successfully finding hydrocarbons. If that were all that were to be done, then one can use the decision-tree diagram of Figure 12.1a to estimate an expected value of

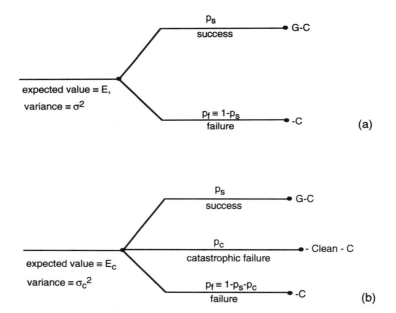

Figure 12.1. Decision tree diagram illustrating: (a) a conventional evaluation scheme for an exploration opportunity; (b) changes brought about by inclusion of a "catastrophic failure" possibility.

$$E = p_s G - C \tag{12.1}$$

and a variance, σ^2, on the expected value of

$$\sigma^2 = p_s p_f G^2 \qquad (12.2)$$

where p_f is the failure probability, $p_f \equiv 1-p_s$. Conventionally one would then estimate the probability, P_+, of obtaining an investment return greater than or equal to zero from

$$P_+ = \sigma^{-1}(2\pi)^{1/2} \int_0^\infty \exp[-(x-E)^2/2\sigma^2]\, dx$$

$$\equiv \pi^{-1/2} \int_{-d}^\infty \exp(-u^2)\, du \qquad (12.3a)$$

where

$$d = -E/(2^{1/2}\sigma) \qquad (12.3b)$$

However, the difficulty arises when a mandated "catastrophic failure" is required to be included, with vanishingly small probability, p_c.

A depiction of inclusion of such a failure in a classical decision-tree is given in figure 12.1b, where "Clean" is the amount one would have to pay for clean-up and equipment loss if the catastrophic failure occurs. And "Clean" is usually very large compared to normal costs, C, of the project and, often, even compared to anticipated gains, G, if the project were to succeed.

In this case, the expected value, E, and the variance, σ^2, in the <u>absence</u> of the catastrophic loss are as given by equations (12.1) and (12.2) respectively, with the probability, P_+, of a positive return again given by equation (12.3b).

Inclusion of the catastrophic loss changes both the expected value E_c and variance, σ_c^2, respectively, to

$$E_c = p_s G - C - p_c \text{Clean} \equiv E - p_c \text{ Clean} \qquad (12.4a)$$

and

$$\sigma_c^2 = p_s G^2 + p_c (\text{Clean})^2 - (p_s G - p_c \text{Clean})^2. \qquad (12.4b)$$

The probability, P_c, of making a return on investment greater than zero is now given by

$$P_c = \sigma_c^{-1} (2\pi)^{-1/2} \int_0^\infty \exp[-(x-E_c)^2/2\sigma_c^2]\, dx$$

$$= \pi^{-1/2} \int_{-D_c}^\infty \exp(-u^2)\, du \qquad (12.5a)$$

where

$$D_c = -E_c/(2^{1/2}\sigma_c) \qquad (12.5b)$$

The point here is that <u>both</u> the expected value <u>and</u> its uncertainty (as measured through the variance) are used to assess the worth of the project. And the relevant comparison to make is between the probability of obtaining a positive return from the project both in the absence and presence of the "catastrophic failure" zone. Consider then, for illustrative purposes, the numerical example sketched in figures 12.2a and 12.2b, both in the absence (figure 12.2a) and presence (figure 12.2b) of the catastrophic loss scenario.

For the assessment in the <u>absence</u> of the catastrophic failure scenario one has, from figure 12.2a, that

$$E = 60, \quad \sigma = 0.55 \times 10^3, \quad E/(2^{1/2}\sigma) = 7.7 \times 10^{-2} \qquad (12.6a)$$

while for the assessment <u>including</u> the catastrophic loss scenario one has, from figure 12.2b, that

$$E_c = -40, \quad \sigma_c = 3.19 \times 10^3, \quad E_c/(2^{1/2}\sigma_c) = -0.89 \times 10^{-2}$$
$$(12.6b)$$

Thus, in this case, while the expected value is shifted from positive to negative by the inclusion of the catastrophic option, the standard error in the mean is increased by a factor of 6, suggesting less accuracy in the mean.

Indeed, a calculation of the probability of obtaining a positive return in the absence of the catastrophic option loss yields

$$P_+ = 54.3\% \qquad (12.7a)$$

while including the catastrophic loss yields the positive return probability of

$$P_c = 49.6\% \qquad (12.7b)$$

Thus the catastrophic loss scenario is ameliorated correctly when the inclusion of the variance on the mean is given. In the specific example given just over a 4% drop occurs in the chance of being profitable, so that the small probability of a catastrophic loss is being included correctly - something which is not possible to have, or even visualize, if one operates with only the expected value as a measure of worth of the opportunity.

B. Catastrophic Loss After Oil is Found

A different type of management concern is to argue that even if an opportunity is to be drilled and if it were to find oil, then there is the probability of a catastrophic oil spill with attendant massive clean-up costs. Such arguments are again often used to downgrade good prospects. In this case the decision tree diagram is different than for the prior case. As sketched in figure 12.3a, the concern here is with catastrophic failure <u>only</u> if oil is found.

In this case, with p_n as the probability of no oil spill and with "Clean" the cost of clean-up if an oil spill does occur, from figure 12.3a one can write

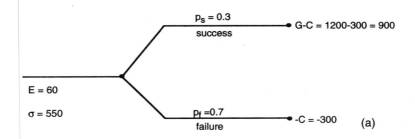

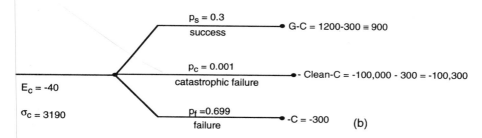

Figure 12.2. Decision tree diagram illustrating: (a) numerical values for figure 12.1a; (b) numerical values for figure 12.1b.

the expected value, E_c, and its variance in the presence of a catastrophic oil spill as

$$E_c = p_s G\text{-}C - p_s(1 - p_n) \text{ Clean} \tag{12.8a}$$

and

$$\sigma_c^2 = p_s\{(1-p_s)G^2 - 2(1-p_s)(1-p_n) G \times \text{Clean}$$
$$+ (1-p_n)(1-p_s(1-p_n))\text{Clean}^2\} \tag{12.8b}$$

Consider, for illustration, the numerical situation depicted in figure 12.3b. Then

$$E_c = -240, \text{ and } \sigma_c = 3.12.10^3 \tag{12.9a}$$

so that

$$E_c /(2^{1/2}\sigma_c) = -5.44 \times 10^{-2} \tag{12.9b}$$

For comparison, note that if the catastrophic oil spill scenario is omitted, then the corresponding values are

$$E = +60, \quad \text{and } \sigma = 0.55.10^3 \tag{12.10a}$$

so that

$$E/(2^{1/2}\sigma) = 7.64 \times 10^{-2}. \tag{12.10b}$$

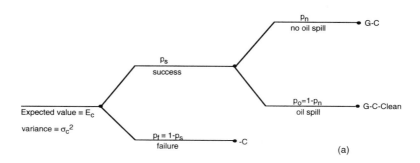

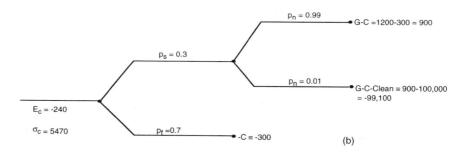

Figure 12.3. Decision tree diagram illustrating: (a) how the inclusion of a catastrophic oil spill after finding oil is included; (b) numerical values for the depiction of figure 12.3a.

In this case, there is a reversal of the expected value (from profitable to non-profitable <u>on average</u>) but there is also an increase in the uncertainty (standard error) on the expected value by an order of magnitude.

The corresponding probabilities of obtaining a positive return are:

(a) in the <u>absence</u> of the catastrophic scenario one has

$$P_+ \cong 54.3\% \qquad (12.11a)$$

(b) in the presence of the catastrophic scenario one has

$$P_c = 48.3\%. \qquad\qquad (12.11b)$$

Thus, inclusion of the variance (into the calculation of the influence of catastrophic oil spill) lowers the chance of making a profit by only 6%. Thus amelioration is again properly taken into account.

C. Conclusions

Natural hazards can lead to catastrophic losses. The costs of platform, pipelines, rigs and other equipment can be measured in millions of dollars. Clean-up cost can be even more. Using the methods described above one can assess the probability of such events which can lead to catastrophic results. Usually these probability are very small. (For example the probability of a medium or strong eruption on the Chirag volcano during the next 30 years is about 10%) (Chapter 4). But, because of the great potential losses, inclusion of such events in economical risk models can swamp completely, in an abusive manner, the characteristics of the opportunity that would exist in the absence of the catastrophic loss conditions if attention is restricted solely to the expected value of the opportunity.

However, such inclusions should also allow for the uncertainty (measured here by the variance or standard error) in the expected value. When such considerations are allowed for, then the abuse of the expected value of a project by such inclusions is not only properly ameliorated but, as also shown here, can be appropriately included in a balanced, rational, objective, reproducible assessment of project worth. The purpose of the case histories given here, and of their numerical illustrations, was to show how such allowances can be made by specific example.

It is strongly recommended that one should avoid using just the expected value as an absolute measure of opportunity worth. Instead one should, more correctly, include the standard error, and also the probability of obtaining a positive return from an opportunity. In this way, low risk, but high potential clean-up cost scenarios are correctly included. And the cases used here have been tailored to maximize the illustration of the way such factors are to be handled.

13 Summary of Potential Natural Hazards

The South Caspian Basin is, without doubt, one of the most hazardous regions of the planet for hydrocarbon exploration. Almost all kinds of geological hazards are present in this small area, which can cause significant problems for operators involved in exploration and development of oil and gas fields.

The reason that so many hazardous phenomena are in this area is due to the unique geological and tectonical setting. As has been mentioned before, the South Caspian Basin, being the result of closure of the Tethys Ocean, was isolated and is still subject to impact of continental plates from different directions. The most active impacts were (at different times) from the Talysh-Vandam and Apsheron-Balkhan directions. Massive compression brought two dominant effects: lateral stress in the sediments and super-rapid sedimentation (Fig. 13.1). Deep earthquakes and Caspian sea level changes can also be classified as tectonic events. In turn, the high sedimentation rate causes overpressure in the formations (which are predominantly shaly) and leads to mud diapirism. In addition, because of the very high sedimentation rates, great thicknesses of unconsolidated sediments are accumulated which can lead to the potential for major landslides. Mud diapirism, which deforms surrounding sediments, also creates local lateral stresses which, in turn, can cause shallow earthquakes.

Mud diapirs are natural storage containers for methane gas. This gas also creates drilling hazards and, of greater danger, the gas forms hydrates near the top parts of a diapir, which can explosively dissociate causing substantial damage during drilling. Rising offshore mud diapirs can form new islands and submarine banks, which are dangerous not only to drilling and production equipment but also to marine traffic.

The surface manifestations of mud diapirs are mud volcanoes which form gryphons and salsas during their quiet phases of activity. (Some gryphons might be formed not in association with mud volcanoes. For example, gryphons are formed in zones of high fracturing and faulting due to overpressure). Those gryphons and salsas are also dangerous. Remember the heights of some gryphon domes are tens of meters, and salsas have the forms of small lakes of tens (sometimes hundreds) of meters in diameter (like Babazanan

335

lake). So imagine what can happen with a platform if one of the legs is lifted or falls by ten or so meters. Gryphons and salsas also emit gas, which is dangerous to human health.

Gas emission also happens during eruptions, which is another fast form of volcanic activity. The emitted gas can flame, another form of hazard. Mud and rock pieces are ejected during eruptions which can also damage equipment. Eruptions also lead to surface fracturing and sometimes to mud island appearances/disappearances. The above is a very schematic overview of the geological hazards in the region.

The hazards investigated in previous chapters are of two sorts: hazards which can be detrimental to equipment (rigs, pipelines, sub-sea completions, supply boats) irrespective of whether one drills or not; and hazards which are either triggered or exacerbated by the very act of drilling.

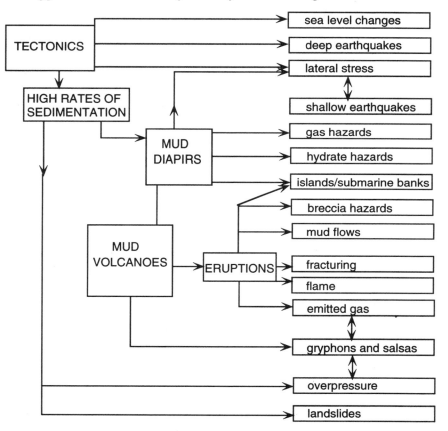

Figure 13.1. Flow chart indicating some of the causes of particular hazards in the South Caspian Basin.

In the category of natural hazards irrespective of drilling, one can list the occurrences of: earthquakes, faults, fractures, mud flows, mud volcanic flames, mud island and submarine banks, breccia, emitted gas, mud ejecta, and hydrate instability. In the category of natural hazards initiated or exacerbated by drilling one can list: overpressure, horizontal stress, hydrates, mud diapir, and dynamical instability.

In all cases the problem is to find some measure for categorizing the likely occurrence of a worst case hazard, and/or of estimating the corresponding strength of the hazard for causing damage.

Every company, before starting any kind of drilling work, usually performs a geological hazard estimate in its area. For that reason, high resolution shallow seismic and sonic surveys are performed. These surveys are useful in terms of sea-bottom investigations. The sonic survey shows the positions of mud volcanoes, gryphons, previous and paleo mud flows, hydrates, ground topography, etc. So, with high probability, the locations of <u>possible</u> centers of hazardous events are located. On the other hand, such surveys do not answer the question: what is the probability of such hazardous events, and what will happen if such events occur? The idea of this book has been to exhibit methods and procedures, and also to provide some examples of how to handle such problems.

When copious, high quality, statistical data are available then there is at least the ability to make some assessment of both the occurrence rate of a hazard together with its magnitude. However, in most situations such data are available only for onshore records, and one already then has an assumption of inference built in when one extrapolates to the offshore regime. In other situations either the data are not available or are sparse and of uncertain quality. In such cases one must not only use the available data with misgivings, but one must also attempt to understand and bracket magnitudes by performing quantitative model calculations.

In addition, many phenomena in the offshore submarine regime either do not occur under aeolian onshore conditions (such as hydrates) or are sufficiently differently behaved in marine conditions (such as mud turbidite flows), that extrapolation to the offshore regime of onshore statistics, even if copious, is not a meaningful exercise.

Two type of hazards specific to the South Caspian Basin region have not been mentioned: first is Caspian Sea level changes, and second is the problem of submarine landslides. The first problem is important because, after almost a hundred years of sea level dropping, the sea-level started to rise rapidly beginning from the end of the 1970's and, during the last 20 years, has increased by almost 2 meters. That rise has brought a lot of problems to the platforms, especially old ones, like Oil Rocks or Sandy Islands (Neft Dashlary and Gum Adasy). The data for sea level profiles with time can be found almost anywhere. There is a published and well-known sea level curve (Dadashev et al., 1995). But dynamical modeling of the processes causing the sea level changes is so complex and includes so many aspects of tectonics, sedimentology, hydrology and climatology, that we did not consider those investigations here in the absence of copious and necessary data.

Another type of hazard not considered in this book is landslides, which are typical of the shelf zone of the region, especially in and around the Apsheron-Balkhan uplift. But the positions of the hazardous zones can be defined by ground sampling and by sonic/seismic surveys. The dynamical scales of such events, i.e., how far submarine ground can slide, could be assessed quantitatively using the MOSED3D program, just as for the submarine mud flows.

Despite these limitations, it is important to provide at least some statements about potential natural hazards in the South Caspian Basin. It would seem that there are copious hazards that one must either guard against or attempt to ameliorate by careful evaluation of rig-siting, drilling conditions, and sub-sea completion equipment locations.

Even such an effort is fraught with difficulty because of the lack of a good offshore data base from which to estimate occurrence rates of a particular hazard or groups of hazards.

Nevertheless, both the statistical measures and the quantitative calculation estimates of hazards presented here are strongly indicative of the need to be extremely cautious in planning and authorizing hydrocarbon exploration endeavors in the South Caspian Basin. And it is this point that has been the main thrust of the efforts presented to assess the potential impact of the various natural hazards that can, and do, occur in the onshore and offshore South Caspian Basin.

REFERENCES

Abasov, M.T., Azimov, E.Kh., Aliyarov, R.Yu., and Ahmedov, A.M., 1991, Theory and practice of geological-geophysical studies and offshore petroleum field development (exemplified by the South Caspian Basin), v. 1, (in Russian): Baku, Elm, 428 pps.

Abikh, G.V., 1939, On the islands occurring in the Caspian Sea and information on the theory of mud volcanoes in the Caspian area. Trudy Instituta, AzFAN, Vol. 12 (in Russian).

Abramovich, M.V., 1959, Exploration of gas accumulations in Azerbaijan related to mud volcanoes. Geologiya Nefti i Gaza, No. 11 (in Russian).

Agabekov, M.G., and Bagirzadeh, F.M., 1948, Eruption of the Hamamdag Mud Volcano, Doklady Akademii Nauk Azerbaijana, vol. IV, No. 11, pp. 477-479.

Agalarova, E.B., 1969, Characteristics of the orientation of stresses acting at the foci of earthquakes in the Apsheron Peninsula and Caspian Sea, Fizika Zemli, No. 7, (in Russian).

Agamirzoev, R.A., 1976, The seismic regionalization of the Azerbaijan SSR, in *Seismotectonics of some regions of the USSR*, Moscow, Nauka, (in Russian).

Agamirzoev, R.A., and Gyul', E.K., 1972, Seismotectonics of the Apsheron area and the Lower Kura Basin, Baku, Elm, (in Russian).

Agamirzoev, R.A., and Gyul', E.K., 1973a, Regional Seismotectonics of the Caspian Sea, in *Seismogenic structures and seismic dislocations*, Moscow, VNIIgeofizika, (in Russian).

Agamirzoev, R.A., and Gyul', E.K., 1973b, The Kura and Caspian earthquakes, Izv. AN AzerbSSR, Ser. Nauk i Zemle, No. 2, (in Russian).

Agamirzoev, R.A., and Gyul', E.K., 1983, Seismotectonics of the Caspian Sea and the characteristics of seismogenic zones, in *Geological and geomorphological studies of the Caspian Sea*, Moscow, Nauka, (in Russian).

Allen, J.R.L., 1985, *Principles of Physical Sedimentology*, George Allen and Unwin, London, 272 p.

Azizbekov Sh.A. and Yakubov, A.A., 1938, "New Island on the Caspian Sea". Izvestiya of Azerbaijan branch of the USSR Academy of Sciences, N1 pp. 23-25, 33.

Bagir-Zadeh, F.M., Kerimov, K.M., and Salayev, S.H., 1988a, The deep structure and oil-gas presence of the South Caspian Mega-Trough. Baku, Azerneshr, 304 pp.

Bagir-Zadeh, F.M., Kerimov, K.M., and Salayev, S.H., 1988b, Subsurface structure and petroleum content of the South Caspian megadepression (in Russian): Baku, Azerneshr, 304p.

Bagirov, E., and Lerche, I., 1996, Quantitative Dynamic Modeling of Mud Diapirism: Evolution of the Vezirov Diapir, in *Evolution of the South Caspian Basin: Geologic Risks and Probable Hazards*, (eds. I. Lerche, E. Bagirov, R. Nadirov, M. Tagiyev and I. Guliev), Azerbaijan Academy of Sciences, Baku, pp. 574-625.

Bagirov, E. and Lerche, I., 1997a, Gas-charging of mud diapirs and fault leakage, in *South Caspian Basin: Stratigraphy, Geochemistry and Risk Analysis*, Azerbaijan Academy of Sciences, Baku, pp. 359-418.

Bagirov, E., and Lerche, I., 1997b, Oil Field Statistics of Azerbaijan: Offshore and Onshore Report, Columbia, SC, 675 p.

Bagirov, E., and Lerche, I., 1998a, Potential Oil-Field Discoveries for Onshore and Offshore Azerbaijan. Marine & Petroleum Geology 15, 11-19.

Bagirov, E., and Lerche, I., 1998b, Flame Hazards in the South Caspian, Energy Exploration and Exploitation 16, 373-400.

Bagirov, E., and Nadirov, R.S., 1995, A method for definition of the power of a mud volcanic eruption. Neft i Gas, No. 2.

Bagirov, E., Nadirov, R., and Lerche, I., 1996, Flaming Eruptions and Ejections from Mud Volcanoes in Azerbaijan: Historical and Statistical Risk Assessment, in *Evolution of South Caspian Basin: Geological Risks and Probable Hazards*, Azerbaijan Academy of Sciences, Baku, 475 p.

Bagirov, E., Nadirov, R., and Lerche, I., 1996a, Flaming Eruptions and Ejections from Mud Volcanoes in Azerbaijan: Statistical Risk Assessment from the Historical Records, Energy Explor. Exploit. 14, 535-584.

Bagirov, E., Nadirov, R.S., and Lerche, I., 1996b, Earthquakes, Mud Volcano Eruptions, and Fracture Formation Hazards in the South Caspian Basin: Statistical Inference from the Historical Record, in *Evolution of the South Caspian Basin: Geologic Risks and Probable Hazards*, Azerbaijan Academy of Sciences, Baku, 625 p.

Bagirov, E. Nadirov, R. and Lerche, I., 1997, Dynamical, Thermal and Hydrocarbon Evolution Across a North-South Section of the South Caspian Basin, in *South Caspian Basin: Stratigraphy, Geochemistry and Risk Analysis*, Azerbaijan Academy of Sciences, Baku.

Bagnold, R.A., 1966, An approach to the sediment transport problem from general physics, Professional Paper, U.S. Geological Survey, No. 422-I.

Baker, P.E., 1972, Experiments on Hydrocarbon Gas Hydrates in Unconsolidated Sand, in *Natural Gases in Marine Sediments*, (ed. I.R. Kaplan), Plenum Press, New York, p. 227-234.

Bitzer, K. and Pflug, R., 1990, DEPO3D: A three-dimensional model for simulating clastic sedimentation and isostatic compensation in sedimentary basins, in *Quantitative Dynamic Stratigraphy*, (ed. T.A. Cross), Prentice Hall, Englewood-Cliffs, p. 335-348.

Bour, O., Lerche, I., and Grauls, D., 1995, Quantitative model of very high fluid pressure - the possible role of lateral stress. Terra Nova 7, 68-79.

Bradbury, R., 1957, *Fahrenheit 451*, Ace Publishing Co., New York.

Bredehoeft, J.D., Djevanshir, R.D., and Belitz, K.R., 1988, Lateral fluid flow in a compacting sand-shale sequence: South Caspian Basin. Am. Assoc. Pet. Geol., Bull, 72, 416-424.

Brown, J.M., 1990, The Nature and Hydrogeologic Significance of Mud Diapirs and Diatremes for Accretionary Systems, Journal of Geophysical Research, 95, 8969-8982.

Brown, J.M., and Westbrook, G.K., 1988, Mud Diapirism and subcretion in the Barbados Ridge Complex, Tectonics, 7, 613-640.

Buniat-Zade, Z.A., and Gorin, V.A., 1968, "About one eruption of gas-oil Volcano Alyat-Cape (March 20, 1967)", Doklady Akademii Nauk Azerbaijana, vol. XXIV, N9, pp. 29-34.

Buryakovsky, L.A. and Djevanshir, R.D., 1983, Filtration and screening properties of clay cap rocks in the zones of anomalously high pore pressures (in Russian). Akad. Nauk. Azerbaijan SSR Izvestiya, Seriya Nauk o Zemle, No. 1, 18-24.

Cao, S., and Lerche, I. 1987, Geohistory, thermal history and hydrocarbon generation history of the nothern North Sea basin, Energy Exploration & Exploitation, 5, 315-355.

Cao, S. and Lerche, I., 1994, A Quantitative Model of Dynamical Sediment Deposition and Erosion in Three Dimensions. Computers and Geoscience, 20, 635-663.

Cao, S., Glezen, W.H., and Lerche, I., 1986, Fluid flow, hydrocarbon generation and migration: A quantitative model of dynamical evolution in sedimentary basins: Proceedings of the Offshore Technology Conference (Houston, TX) paper 5182, vol. 2, p. 267-276.

Claypool, G.E., and Kaplan, I.R., 1972, The Origin and Distribution of Methane in Marine Sediments, in *Natural Gases in Marine Sediments* (ed. I.R. Kaplan), Plenum Press, New York, pp. 99-139.

Dadashev, F.G., 1963, Hydrocarbon gases from the mud volcanoes of Azerbaijan; Azerneshr., Baku (in Russian).

Dadashev, F.G. Guseynov, R.A., and Aliyev, A.I., 1995, Map of Mud Volcanoes of the Caspian Sea (Explanatory Notes). Geological Institute of Azerbaijan Academy of Sciences, Baku, 20 pp.

Dadashev, F.G., and Mekhtiyev, A.K., 1975, Mud volcanoes of the Caspian Sea, Izvestiya AN Azerb. SSR, Serya Nauk o Zemle (Earth Sciences Branch), No. 5, 26-32 (in Russian).

Enron Corporation, 1993, The Outlook for Natural Gas, 16 p.

Feller, W., 1971, *An Introduction to Probability Theory and Its Applications*, Vol. 1, John Wiley and Sons, New York.

Gambarov, J.G, Jafarov, Z.F., Mamedov, P.Z., and Shykhaliev, J.A., 1993, The seismostratigraphic and structural-formational analysis of the regional profiles (cross-sections) data of the South Caspian Mega-Trough, (p. 94-116), in *The Structural-formational and seismostratigraphical investigations of the South Caspian Mega-Trough sedimentary cover*. Azniigeophizika, Baku, 128 pp.

Garde, R.J., and Ranga Raju, K.G., 1978, *Mechanisms of Sediment Transportation and Alluvial Stream Problems*, Wiley Eastern Limited, New Delhi, 483 p.

Gasanov, I.S., Ismail-Zadeh, T.A., Rahimkhanov, F.G., and Kerimov, K.M., 1988, Atlas of petroleum bearing and prospective structures of the Caspian Sea (in Russian): Leningrad, Ministry of Geology, USSR.

Gilev, K.V., 1872, Information on an oil-field in Baku province and Kaitag-Tabasaran county. In: "Collection of information about Caucasus", vol. 2, Tiflis, pp. 51-72 (in Russian).

Ginsburg, G.D., Guseynov, R.A. Dadashev, A.A., Ivanova, G.A., Kazantsev, S.A., Solov'yev, V.A., Telepnev, E.V., Askeri-Nasirov, R. Ye., Yesikov, A.D., Mal'tseva, V.I., Mashirov, Yu.G., and Shabayeva, I. Yu., 1992, Gas Hydrates of the South Caspian, International Geology Review 34, 765-782.

Golubyatnikov, D.V., 1904, Eruption of Otmanbozdag Mud Volcano on November 10, 1904, - Kaspii, (in Russian).

Gorin, V.A., 1950, Eruption of Bolshoi Kanizadag Mud Volcano, Doklady Akademii Nauk Azerbaijana, vol. VI, No. 7.

Gradshteyn, I.S., and Ryzhik, I.M., 1965, *Table of Integrals, Series and Products*, Academic Press, New York.

Gubkin, I.M., 1934, *Tectonics of the South-East part of the Caucasus relating to oil and gas presence in this area*. Mineral, Geologic and Oil Publishing House, Moscow, 51 p. (in Russian).

342

Guidish, T.M., Kendall, C.G.St.C., Lerche, I., Toth, D.J., and Yarzab, R.F., 1985, Basin evolution using burial history calculations: an overview: AAPG Bull., 69, 92-105.

Guliev, I.S., 1995, Private Communication.

Guliev, I.S., 1996, Model of Mud Volcanism, in *Evolution of the South Caspian Basin: Geologic Risks and Probable Hazards*, (eds. I. Lerche, E. Bagirov, R. Nadirov, M. Tagiyev and I. Guliev), Azerbaijan Academy of Sciences, Baku, pp. 549-573.

Guliev, I.S., 1997, Model of Mud Volcanism. Chapter 6 of *Evolution of the South Caspian Basin: Geological Risks and Probable Hazards*: Azerbaijan Academy of Sciences, (eds. I. Lerche, etc.) (second edition), pp. 506-530.

Guliev, I.S., Frantsu, Yu., Feizullayev, A.A., Muller, P., and Mamedova, S.A., 1991, Geological-geochemical features of petroleum formation in alpine intermontane depressions (in Russian). Geokhimiya, No. 1: 148-156.

Hitchon, B., 1972, Occurrence of Natural Gas Hydrates in Sedimentary Basins, in Natural Gases in Marine Sediments (ed. I.R. Kaplan), Plenum Press, New York, pps. 195-225.

Hobbs, B.E., Means, W.D., and Williams, P.F., 1976, *An Outline of Structural Geology*. John Wiley & Sons, Singapore, 571 p.

Hot Ice, 1991, "Antenna" series of BBC-2. Produced by GEOFILMS Ltd.

Ismailzadeh, I.G., 1954, "Offshore mud volcanoes", Priroda, N. 11, pp. 94-95 (in Russian).

Ivanov, V.V, and Guliev, I.S., 1986, An attempt at physical-chemical modeling of mud volcanism. Bull. Moscow Society of Naturalists, Geol. Series, Vol. 61, No. 1 (in Russian).

Ivanov, V.V., and Guliev, I.S., 1987, Physical-chemical model for mud volcanism, in *Problems of Oil and Gas in the Caucasus*. Moscow, Nauka (in Russian).

Jaeger, J.C. and Cook, N.G.W., 1976, *Fundamentals of Rock Mechanics*, Chapman and Hall, London, 392 pp.

Kaplan, I.R. (ed.), 1972, *Natural Gases in Marine Sediments*, Plenum Press, New York.

Kastrulin, N.S., 1979, Mud volcanism as an index of prediction in oil and gas reserves evaluation. Doklady AN Azerb. SSR, v. 35, No. 6, 70-74 (in Russian).

Khalilbeili, Ch.A., and Gasanov, A.N., 1960, "Eruption on Kumani Island", Azerbaijanskoye Neftyanoye Khozyajstvo, No. 10, pp. 9-11 (in Russian).

Korchagina, Yu.I., Guliev, I.S., and Zeinalova, K.S., 1988, Petroleum generative potential of deep-seated Mesozoic-Cenozoic sediments of the South Caspian basin: Problems of petroleum content of the Caucasus region (in Russian), Moscow, Nauka, p.35-41.

Kovalevskiy, C.A., 1940, *Mud volcanoes of the South Caspian (Azerbaijan and Turkmenistan)*, Azgostoptehizdat, Baku, (in Russian).

Kremenetzky, A.A., 1988, Petrological study of samples from the section of Saatly super-deep well (in Russian). Moscow, Nauka.

Kremenetzky, A.A., Lapidus, A.V., and Skrjabin, V.J., 1990, Geological and geochemistry methods for the forecast of the deep-laid mineral resources. Moscow, Nauka, 223 pp. (in Russian).

Kropotkin, P.I., and Valyayev, M.B., 1981, Geodynamics of mud volcano activity (with connection to oil and gas presence), in *Geological and Geochemical backgrounds of oil and gas exploration*. Naukova Dumka, Kiev, (in Russian).

Kuliev, F.T., 1979, Equation for the macroseismic field of Azerbaijan. Seis. Bull. of the Caucasus, (in Russian).

Lerche, I., 1990a, *Basin Analysis: Quantitative Methods*, Vol. 1, Academic Press, San Diego, 562 pp.

Lerche, I., 1990b, *Basin Analysis: Quantitative Methods*, Vol. 2, Academic Press, San Diego, 570 pp.

Lerche, I. and Bagirov, E., 1993/94, Hydrate Hazards in the South Caspian Basin. Izvestiya Akademii Nauk Azerbaijana. Seriya Nauk o Zemle N°1-6, p. 116-124 (in Russian).

Lerche, I., Bagirov, E., Nadirov, R., Tagiyev, M., and Guliev, I., 1996, *Evolution of the South Caspian Basin: Geologic Risks and Probable Hazards*. Nafta-Press, Azerbaijan Academy of Sciences, Baku, Azerbaijan, 625 p.

Lerche, I. and O'Brien, J.J., 1987a, *Dynamical Geology of Salt and Related Structures*, Academic Press, Orlando, 832 pp.

Lerche, I. and O'Brien, J.J., 1987b, Modeling of Buoyant Salt Diapirism, in *Dynamical Geology of Salt and Related Structures*, Academic Press, Inc., 129-162.

Lerche, I., Yarzab, R.F., and Kendall, C.G.St., 1984, Determination of paleoheat flux from vitrinite reflectance data, AAPG Bulletin, 68, 1704-1717.

Lowrie, A., 1997, Hydrates in the Gulf of Mexico and Eastern United States. Paper presented at the Gulf Coast Association of Geological Societies meeting, Houston, October 1997.

Malinovskii, N.V., 1938, Seisms accompanying mud eruptions. Trudy AzFAN SSSR. Ser. fiziko-khimicheskaya, Vol. 3/38 (in Russian).

Middleton, G.V., 1966, Experiments on density and turbidity currents. II. Uniform Flow of density currents, Can. J. Earth Sci., 3, 627-637.

Miller, S.L., 1972, The Nature and Occurrence of Clathrate Hydrates, in *Natural Gases in Marine Sediments* (ed. I. R. Kaplan), Plenum Press, New York, pp. 151-177.

Mohr, O., 1900, Welche Umstande bedingen die Elastizitatsgrenze und den Bruch eines Materials? Z. Ver. dt. Ing. 44: 1524-1572.

Nadirov, R.S., 1983, On the origin and the nature of the Talysh-Vandam uplift in Azerbaijan, Proceedings of the Conference on The Earth's Crust Outgassing and Geotectonics, Moscow, Nauka, p. 68-73 (in Russian).

Nadirov, R.S., 1985, The geological structure and oil and gas presence of the Talysh-Vandam gravitational maximum area. Dissertation thesis summary (in Russian), Baku, 30 pp.

Nadirov, R.S., Bagirov, E., Tagiyev, M., and Lerche, I., 1997, Flexural plate subsidence, sedimentation rates, and structural development of the super-deep South Caspian Basin. Marine and Petroleum Geology 14, 383-400.

Nadirov, S.G., and Zeinalov, M.M., 1958, "A New Eruption of Bozdag (Kobiiski) Mud Volcano", Azerbaijanskoye Neftanoye Khozyaistvo, N3, pp. 9-10 (in Russian).

O'Brien, J.J. and Lerche, I., 1987, Heat Flow and thermal maturation near salt diapirs, in *Dynamical Geology of Salt and Related Structures*, (eds. I. Lerche and J. J. O'Brien), Academic Press, Orlando, pp. 711-750.

O'Brien, J.J. and Lerche, I., 1987, Modeling of the deformation and faulting of formations overlying an uprising salt dome, in *Dynamical Geology of Salt and Related Structures*, (eds. I. Lerche and J. J. O'Brien), Academic Press, Orlando, p. 419-455.

Panakhi, B.M., 1988, Seismotectonic territories of the Caspian Sea, Author's abstract for a Candidate's dissertation, Tbilisi, Georgia, (in Russian).

Panakhi, B.M., and Kasparov, V.A., 1988, Problems of the seismic region in the Caspian Sea. Izv. AN. AzerbSSR., Serya Nauk o Zemle, No. 1, (in Russian).

Parker, E.N., 1978, *Cosmical Magnetic Fields*, Oxford University Press, Oxford, 937 p.

Petersen, K. and Lerche, I., 1995, *Salt and Sediment Dynamics* , CRC Press, Boca Raton, 322 pp.

Potapov, I.I., 1935, An eruption of Lokbatan Mud Volcano on February 23, 1935, Azerbaijanskoye Neftyanoye Khozyaystvo, N6, pp. 27-30 (in Russian).

Putkaradze, A.L., Mamedov, M.K., and Mustafayev, I.S., 1954, "Eruption of the Buzovninskay Sopka Volcano", Azerbaijanskoye Neftyanoye Khozyaystvo, N4, pp. 4-9.

Rakhmanov, R.R., 1987, *Mud volcanoes and their significance in oil and gas prediction.* Nedra, Moscow, 176 p. (in Russian).

Richter, C.F., 1958, *Elementary Seismology*, W.H. Freeman & Co., San Francisco.

Riznichenko, Yu V. (ed.) 1960, Methods for the detailed study of seismicity. Trudy IFZ AN SSSR. No. 9, (in Russian).

Riznichenko, Yu V., 1985, Selected works, problems in seismology, Moscow, Nauka, (in Russian).

Salakhov, A.S., 1985, The Early- and Middle Jurassic volcanism of the Saatly-Kurdamir uplift according to the Saatly superdeep well section investigation. Dissertation thesis summary. Baku, 25 pp. (in Russian).

Sclater, T.G. and Christie, P.A.F., 1980, Continental stretching: an exploration of the post-mid-Cretaceous subsidence of the Central North Sea basin: Journal of Geophys. Res., 85, 3711-3739.

Shebebalin, N.V., 1955, On the relationship between the energy and focal depth of earthquakes, Izv. ANSSSR, Ser. Geofiz., No. 4, (in Russian).

Shebebalin, N.V., 1968, Methods for using engineering and seismological data for establishing seismic regions, in *Seismic Regions of the USSR*, Nauka, Moscow, (in Russian).

Sjögren, Hj. Der Ausbruch des Schlammvulcans Lok-Batan am Kaspischen Meere vom 5 Januar 1887. - "Jahrb. der Wien. Geol. Reichsanstalt", 1888, Bd. 37, Heft 2, p. 237-244.

Sjögren, Hj. Üeber die Thätigkeit der Schlammvulkane in der Kaspischen Region Wahrend der Jahre 1885-1887. - Papers of Imperial Minerological Society, vol. 2, part 24, pp. 1-22.

Sjögren, Hj. Om narg a Slammvulkanubrott im Kaspiska regionen under jahre 1892-1896. "Geol. Förein. Förhand", 1897, bd. 19, No. 176, pp. 91-105.

Solheim, A., and Elverhøi, A., 1993, Gas-related seafloor craters in the Barents Sea. Geo-Marine Letters 21, 12-19.

Sultanov, A.D., and Agabekov, M.G., 1959, "Makarov Bank mud volcano eruption", Doklady Akademii Nauk Azerbaijanskoi SSR, vol. XV, 1959, No. 2 (in Russian) pp. 143-146.

Sultanov, A.D., and Dadashev, F.G., 1962, Mud volcano eruption on Duvanny Island - Izvestiya Akademii Nauk Azerbaijanskoj SSR, Seriya Geologo-geographicheskikh nauk i nefti, No. 3, p. 73-81 (in Russian).

Sultanov, R.G., Khalilbeili, Ch.A., and Suleimanov, A.I., 1967, New data on the geological construction of Bulla Island (Baku Archipelayea), "Uchenyie Zapiski" of Azerbaijan State University. Division of Geology and Geography, N2, pp. 44-52 (in Russian).

Tagiyev, M.F., Nadirov, R.S., Bagirov, E.B., and Lerche, I., 1996, Geohistory, Thermal History and Hydrocarbon Generation History of the north-west South Caspian Basin, in *Evolution of the South Caspian Basin: Geologic Risks and Probable Hazards* (eds. I. Lerche, E. Bagirov, R. Nadirov, M. Tagiyev and I. Guliev), Azerbaijan Academy of Sciences, Baku, pp. 50-106.

Taylor, G.I., 1953, *Collected Papers of Sir G.I. Taylor*, Vol. 3, Oxford University Press, Oxford.

Tetzlaff, D.M., 1990, SEDO: A simple clastic sedimentation program for use in training and education, in *Quantitative Dynamic Stratigraphy*, (ed. T.A. Cross), Prentice Hall, Englewood-Cliffs, p. 401-415.

The Bermuda Triangle, 1992, Equinox Science Series-Channel Four Television, Produced by Geofilms Ltd..

Tissot, B., and Welte, D.H., 1978, *Petroleum Formation and Occurrence*, Springer-Verlag, New-York, 538 pp.

U. S. Geological Survey Professional Paper 1570 (ed. D. G. Howell), 1993, The Future of Energy Gases, United States Government Printing Office, Washington, D.C., 890 pp.

Ungerer, P., Bessis, F., Chenet, P.Y., Durand, B., Nogaret, E., Chiarelli, A., Oudin, J.L., and Perrin, J.F., 1984, Geochemical and geological methods in oil exploration, principles and practical examples, AAPG Memoir 35, 53-77.

Valyaev, B.M., Grinchenko, Yu.I., and Erokhin, V.E., 1985, Isotopic aspects of mud volcanoes, Litologiya i Poleznye Iskopayemye, No. 1.

Welte, D.H. and Yukler, M. A., 1981, Petroleum origin and accumulation in basin evolution - a quantitative model, AAPG Bull., 65, 1387-1396.

Yakubov, A.A., Ali-zade, A.A., Rakhmanov, R.R., and Mamedov, Yu, G., 1974, Catalog of recorded eruptions of mud volcanoes in Azerbaijan in the period 1810-1974, Baku, Azerbaijan Academy of Sciences Report (in Russian).

Yakubov, A.A., Dadashev, F.G., and Magerramova, F.S., 1965, Eruption of Ayrantekyan Mud Volcano, Doklady Akademii Nauk Azerbaijana, vol. XXI, N2, pp. 33-37.

Yakubov, A.A., Dadashev, F.G., and Zeinalov, A.M., 1970a, About new eruptions of mud volcanoes on the south-east part of the Greater Caucasus, Elm, Baku.

Yakubov, A.A., Gadjiyev, Y.A., Matanov, F.A., and Atakishiyev, I.S., 1970b, "Eruption of Kelany Mud Volcano", Doklady Akademii Nauk Azerbaijana, vol. XXVI, No. 5, pp. 55-60.

Yakubov, A.A., Grigoryanc, B.V., and Aliyev, A.A., 1980, *Mud volcanism in the Soviet Union and its connection with oil and gas presence*. Elm, Baku, 165 p. (in Russian).

Yakubov, A.A., Matanov, F.A. and Atakishiyev, I.S, 1972, Peculiarity of Eruption of Cheildag Mud Volcano, Izestiya Akademii Nauk Azerbaijana, N2, pp. 3-9 (in Russian).

Yakubov, A.A. and Putkaradze, A.L., 1950, Eruption of Bolshoi Kanizadag mud volcano, Azerbaijanskoye Neftyanoye Khozyaystvo, No. 9, pp. 1-4 (in Russian).

Yakubov, A.A. and Putkaradze, A.L., 1951, "The eruption of Kumani Island", Azerbaijanskoye Neftyanoye Khozyaystvo, N2, pp. 3-5 (in Russian).

Yakubov, A.A. and Salayev, S.G., 1953, Eruption of Bozdag Kobiiski Mud Volcano, Azerbaijanskoye Neftyanoye Khozyaystvo, N12, pp. 3-4 (in Russian).

Zeinalov, M.M., Kagramanov, K.C., Zeinalov, A.M. and Magerramova, F.S., 1969, Eruption of Melikchobanly Mud Volcano, Doklady Academii Nauk Azerbaijana, vol. XXV, N5, pp. 43-48 (in Russian).

Zhemerev, V.S., 1954, Eruption of Lokbatan mud volcano, Azerbaijanskoye Neftyanoye Khozyaystvo, N11, pp. 30-31 (in Russian).

Zuber, S.R., 1923, "Study of the sea bottom to the North from Absheron Peninsula", Azerbaijanskoye Neftyanoye Khozyaystvo, N1, p. 126 (in Russian).

INDEX